Options for

Remote Monitoring and Control of Small Drinking Water Facilities

Albert J. Pollack
Abraham S. C. Chen
Roy C. Haught
James A. Goodrich

BATTELLE PRESS
Columbus • Richland

Library of Congress Cataloging-in-Publication Data

Options for remote monitoring and control of small drinking water facilities / Albert J. Pollack ... [et al.].
p. cm.
Includes bibliographical references and index.
ISBN 1-57477-072-1 (alk. paper)
1. Waterworks--Automation. 2. Water treatment plants--Automation.
3. Information storage and retrieval systems--Water-supply.
I. Pollack, Albert J., 1949-
TD487.068 1999
628.1'62--dc21 99-14431
CIP

DISCLAIMER

The vendors and products identified in this book are primarily for information purposes only. Mention of these vendors and suppliers does not constitute endorsement or recommendation for use by Battelle. This book contains information obtained from authentic and highly regarded sources. Reprinted material is quoted with permission, and sources are indicated. A wide variety of references are listed. Reasonable efforts have been made to publish reliable data and information, but the authors and the publisher cannot assume responsibility for the validity of all materials or the consequences of their use. Reported specifications, performance characteristics, and costs of the components identified in this book are taken from published manufacturers' catalogs and specification sheets, and from verbal conversations and quotes. Actual operational characteristics and costs have not been verified by the authors.

Printed in the United States of America

Battelle Press
505 King Avenue
Columbus, Ohio 43201-2693
614-424-6393; 1-800-451-3543
Fax: 614-424-3819
Homepage: http://www.battelle.org/bookstore
E-mail: press@battelle.org

Foreword

With recent advancements in both electronic sensors/monitors and computerized data acquisition/control capabilities, many routine manual activities associated with production or service operations are now being performed by automated systems. Once data have been collected and stored, current communication options make it possible to retrieve data without having to physically visit the site. This transmission can be achieved by employing technologies that range from hard-wired phone lines to wireless telemetry via radio frequency transmitters or low earth orbit satellites.

A large number of small communities have limited financial resources for drinking water system management and operation. Costs for these services can be a barrier to providing adequate supplies of reliable and safe drinking water. Sparse populations, compounded with a low tax base, may make comprehensive monitoring and maintenance of small water treatment systems prohibitively expensive. As a result, small systems are responsible for the majority of monitoring and reporting violations under the Safe Drinking Water Act Amendments of 1996. The availability of sensors, data acquisition systems, and transmission components, when properly integrated, could permit the monitoring and control of small drinking water facilities. The United States Congress recognized this situation and directed the United States Environmental Protection Agency (U.S. EPA) to identify affordable technologies for public water systems serving all small communities.

This book is a compilation of typical small drinking water facility operations and monitoring requirements with commercial off-the-shelf technologies provided that can assist the reader in selecting components for compliance reporting. The material in this book will enable individuals to better understand the monitoring devices that historically have been used while identifying emerging technologies in monitoring, data acquisition, and data transmission. Remote monitoring, data collection and storage, and advancements in telemetry will enable one qualified operator to review the operation of several treatment systems using real-time data transmission from the treatment facilities to a central control center. Manual maintenance and testing activities will never be removed completely from the operation of small community drinking water facilities. However, the implementation of automated monitoring and remote system control should reduce the costs associated with daily site visits, bring facilities into monitoring compliance, and most importantly identify situations that could compromise making safe potable water available for all citizens.

Because of ongoing advancements in the technology associated with electronic monitoring, data acquisition, and data transmission components, it is not possible to identify all of the options and vendors for the identified technologies. The intention of the authors was to provide an overview of applicable components for remote monitoring and telemetry. It is likely that additional options already exist and that new components will be developed that could be used as alternatives to the mentioned technologies.

Much of this book was prepared originally in a report as a deliverable to the National Risk Management Research Laboratory, Office of Research and Development, of the U.S. EPA under Contract No. 68-C7-0008, Work Assignment No. 1-08. The manuscript has not been subjected to U.S. EPA's review and, therefore, does not necessarily reflect the views of the U.S. EPA, and no official endorsement should be inferred.

The authors would like to acknowledge and thank those individuals who provided input for this book. Drs. Vernon L. Snoeyink and Benito J. Marinas of the University of Illinois at Urbana-Champaign provided small drinking water facility statistical information, summaries of regulatory requirements, and discussions on the amenability of various monitoring and control technologies. Ms. Kristen M. Ruffell, also from the University of Illinois at Urbana-Champaign, coordinated the compilation of the above information and oversaw its transfer to Battelle in a timely fashion. Ms. Julie M. Kramer, Mr. Clifford N. Dodson, Mr. Mark Koenig, and Dr. Vilayanur V. Viswanathan of Battelle compiled and summarized data on commercially available items that could be used to monitor water quality parameters and control system operations as well as provide data acquisition and transfer under a remote telemetry system platform.

Finally, the authors wish to thank Ms. Lynn Copley-Graves, Ms. Loretta E. Bahn, Mr. Dean Sutton, Mr. Victor J. Saylor, and Mr. Sumedha De Silva, who were instrumental in the editorial, text processing, graphics production, and data compilation aspects of this book.

Contents

Appendices

Index

Figures

Tables

Acronyms, Abbreviations, and Symbols

A

AA	activated alumina; atomic absorption
AC	alternating current
A/D	analog to digital
ADC	analog to digital conversion
AFSC	Air Force Space Command
AMA	American Medical Association
ASCII	American Standard Code for Information Interchange
ASTM	American Society for Testing and Materials
ATC	automatic temperature compensation
A.U.	absorbence unit(s)
AWWA	American Water Works Association

B

BFSL	best-fit straight line
BNC	bayonet N connector
bps	bits per second

C

C	control
CAB	Commercial Agriculture Bureau
COTS	commercial-off-the-shelf
CPU	central processing unit
CPVC	chlorinated polyvinyl chloride
CSI	Campbell Scientific, Inc.
CT	product of concentration and t_{10} contact time

D

D	deep; depth
DAC	digital to analog conversion
DAQ	data acquisition
DBP	disinfection by-product
DC	direct current
DDS	direct digital system
DEC	Digital Equipment Corporation
DI	deionized
DKA	double known addition
DMA	direct memory access
DNA	deoxyribonucleic acid
DO	dissolved oxygen
DoD	U.S. Department of Defense
DOS	disk operating system
DPD	N,N-diethyl-*p*-phenylenediamine
DPR	differential pressure range

E

ED	electrodialysis
EDR	electrodialysis reversal
EDTA	ethylenediaminetetraacetic acid
EEPROM	electrically erasable, programmable, read-only memory

ESWTR	Enhanced Surface Water Treatment Rule
ETSI	European Telecommunications Standards Institute

F

FCC	Federal Communications Commission
Fe	iron
FMI	Fluid Metering, Inc.
FS	full scale
FTU	Formazin nephelometric turbidity unit

G

GAC	granular activated carbon
GHz	gigahertz
GLI	GLI International, Inc.
GOES	Geostationary Operational Environmental Satellite
gph	gallon(s) per hour
GPIB	General Purpose Interface Bus
gpm	gallon(s) per minute
GPS	Global Positioning System
GUI	graphic user interface
GW	groundwater
GWDR	Groundwater Disinfection Rule

H

H	height; high
HAA	haloacetic acid
HI	Hanna Instruments, Inc.
HOCl	hypoclorous acid
Hz	hertz

I

IAC	Information Access Company
IC	ion chromatography
ICP	inductively coupled plasma
IHS	Information Handling Systems
I/O	input/output
IOC	inorganic compound
IR	infrared; infrared spectroscopy
ISA	industry standard architecture
ISE	ion-selective electrode
IX	ion exchange

K

kB	kilobyte(s)
kbps	kilobits per second
kHz	kilohertz

L

L	length; long
LCD	liquid crystal display
LCP	liquid crystal polymer
LED	light-emitting diode
LEO	low earth orbit
LSD	light-scattering division

M

mA	milliampere
mbar	millibar
MCL	maximum contaminant level
MCLG	maximum contaminant level goal
MEO	medium earth orbit
MF	microfiltration; million fibers
mg/L	milligrams per liter
MHz	megahertz
Mn	manganese
MPT	male pipe thread
MRDL	maximum residual disinfectant level
mrem	dose to reference man
mS	millisiemens
MWCO	molecular weight cutoff
µS	microsiemens

N

NA	not applicable
NEMA	National Electrical and Mechanical Association
NF	nanofiltration

NIST	National Institute of Standards and Technology
NMRI	Naval Medical Research Institute
NOAA	National Oceanic and Atmospheric Administration
NPT	normal pipe thread
NPTF	normal pipe thread, female
NPTM	normal pipe thread, male
NR	not reported
NRC	National Research Council
NRMRL	National Risk Management Research Laboratory
NTIS	National Technical Information Service
NTU	nephelometric turbidity unit

O

OD	outside diameter
O-E	opto-electronic
ORP	oxidation-reduction potential

P

PAC	powdered activated carbon
PAH	polycyclic aromatic hydrocarbon
PC	personal computer
PCB	polychlorinated biphenyl
PCI	peripheral component interconnect
PCMCIA	Personal Computer Memory Card International Association
PCR	polymerase chain reaction
POE	point-of-entry
POTS	plain old telephone system
POU	point-of-use
ppb	part(s) per billion
ppm	part(s) per million
PPS	Precise Positioning Service
ppt	part(s) per thousand
psi	pound(s) per square inch
psia	pound(s) per square inch absolute
psid	pound(s) per square inch, differential
psig	pound(s) per square inch, gage
PVC	polyvinyl chloride
PVDF	polyvinylidene difluoride

R

RAM	random access memory
RF	radiofrequency
RH	relative humidity
RNA	ribonucleic acid
RO	reverse osmosis
rpm	revolutions per minute
rRNA	ribosomal ribonucleic acid
RTD	resistance temperature detector
RTS	remote telemetry system
RTU	remote telemetry unit

S

SCADA	supervisory control and data acquisition
SDWA	Safe Drinking Water Act
SDWAA	Safe Drinking Water Act Amendments
SMCL	secondary MCL
SOC	synthetic organic compound
SPS	Standard Positioning Service
SRAM	static random access memory
SS	stainless steel
SSU	small subunit
STD	standard
SW	surface water
SWTR	Surface Water Treatment Rule

T

TC	treatment credit
TDG	total dissolved gas
TDS	total dissolved solids
THM	trihalomethane
TISAB	total ionic strength adjusting buffer
TOC	total organic carbon
TPTZ	2,4,6-tripyridyl-*s*-triazine
T.T.	treatment technique
TTL	transistor-transistor logic

U

UF	ultrafiltration
UHF	ultrahigh frequency
U.S. EPA	United States Environmental Protection Agency
UV	ultraviolet

V

VAC	volt(s) alternating current
VDC	volt(s) direct current
VHF	very high frequency
VME	VERSAmodule Eurocard
VOC	volatile organic compound

W

W	watt(s); wide, width
WSWRD	Water Supply and Water Resources Division

1

Introduction

Approximately 190,000 very small and noncommunity drinking water treatment systems in the United States serve transient and nontransient populations of 3,300 people or fewer (Lykins and Clark, 1996). Because these systems are widely dispersed and face financial constraints, it is difficult to maintain extensive manual monitoring to ensure compliance with regulatory mandates. Consequently, these small treatment systems are the most frequent violators of the Safe Drinking Water Act (SDWA) and its Amendments (SDWAA), and thus are responsible for approximately 90% of the monitoring/reporting violations (NRC, 1997).

In an effort to address this problem, the Water Supply and Water Resources Division (WSWRD) of the National Risk Management Research Laboratory (NRMRL), U.S. Environmental Protection Agency (U.S. EPA) has been evaluating the option of utilizing remote telemetry of an "electronic-circuit-rider" that would allow one qualified operator to monitor and control the operation of several small treatment systems from a centrally located computer (Teuschler et al., undated). The use of such a system would optimize the time spent for on-site inspection and maintenance efforts, allowing the operator to visit only the problematic systems as indicated by the remote telemetry system (RTS). The results expected from an appropriately designed and successfully deployed remote monitoring and control system would include enhanced water quality, compliance with existing water quality regulations, and reduced costs for small communities.

This book examines the parameters amenable to remote monitoring and control of small systems; identifies discrepancies between required parameters and the current availability of suitable sensors or controls; and determines needs for additional analytical methods development, remote telemetry research, and system integration. Specifically, this book does the following:

- Reviews and assesses the open literature regarding water services to small communities, and water quality and system operating parameters amenable to small-system remote control and operation.
- Identifies parameters that are required and amenable to remote monitoring and control of small systems.
- Reviews and assesses the open literature regarding the state of the art of data acquisition (DAQ) methods available to integrate monitoring and control components, and of data transmission systems capable of transmitting signals from the RTS.
- Reviews the availability of commercial-off-the-shelf (COTS) sensors, process control devices, supervisory control and DAQ equipment, and remote telemetry system packages.
- Examines research needs for additional analytical sensing and/or remote telemetry methods.

The review performed by the University of Illinois focused on generating a database of water services to small communities and an assessment of the types of water quality and operating parameters amenable to remote sensing and control, and a database of COTS devices, packages, and equipment, as well as their applicability and usefulness in providing automation and remote sensing/control for small treatment systems. Special attention was paid to each component's stability, sensitivity, reliability, and ease of integration into an RTS. Software packages were reviewed to determine their availability for integrating monitoring and control components under a suitable format. Options for data transmission from the RTS were reviewed for their advantages and limitations with respect to interface constraints, cost effectiveness, and ease of use.

1.1 Literature Search Methodology

Three literature searches were performed. The first covered COTS sensor technologies that typically are associated with treatment plant activities for monitoring water quality. The search driver was provided with keywords for both academic journals and trade sources. This two-pronged approach was done so that if sensors for a specific parameter were not commercially available, the literature indicating recent research into the development of such a device would be found. The second search was directed at identifying sources of components for monitoring operational parameters of package water treatment plants. For this search, the focus was on trade magazine databases because devices such as pressure monitors, flowmeters, totalizers, and level sensors were expected to be commercially available. The third and final search covered DAQ and remote telemetry systems. Due to the evolving nature of the electronics industry, both scientific and trade databases were searched. In each of the three searches, keywords that would qualify and focus the search for potential aspects of the RTS were used. A detailed description of the search methodology is presented in Appendix A of this report.

After component sources had been identified, representative manufacturers, suppliers, vendors, and distributors of the items of interest were asked to provide literature on their product line. The technical materials received were reviewed; data sheets were completed for each of the represented technologies; and spreadsheets summarizing the information were prepared.

With the resources available, it was not possible to identify and document every type of sensor, process monitoring component, DAQ system, and telemetry option available as a commercial item. However, the literature search did result in an extensive database that includes predominant, established, and field-tested technologies applicable to the RTS.

1.2 Organization

The body of this book starts with statistics relating to small treatment systems and regulatory requirements applicable to these systems, and the treatment technologies currently used by most small treatment systems and the treatment plants suitable for remote monitoring and control (Chapter 2). Chapter 3 discusses the water quality and system operating parameters amenable to small-system remote control and operation. Chapters 4 and 5 summarize the results of the literature search for COTS availability of sensors and process control devices for water quality and system operation, respectively. Spreadsheets are used to describe various COTS products under each water quality and system operating parameter. Chapters 6 and 7 provide an overview of DAQ and data transmission technologies, respectively. These chapters also discuss system integration/interface between sensors, controls, and various telemetry options. Spreadsheets are used to compare various COTS products. Chapter 8 presents selection criteria that may aid in the selection of individual devices, components, DAQ systems, and remote telemetry options for inclusion in an RTS. Chapter 9 identifies the research needs for additional analytical/sensing and/or remote telemetry methods. Appendices include a detailed description of the literature review strategy (Appendix A), as well as individual data sheets of all COTS products obtained from various vendors and manufacturers for water quality monitoring and control (Appendix B), system operation monitoring and control (Appendix C), DAQ (Appendix D), and remote telemetry (Appendix E).

1.3 References

Lykins Jr., B.W.; and Clark, R.M. (1996), Water Treatment Technologies for Small Communities, presented at the 2nd EPA National Drinking Water Treatment Technology Transfer Workshop, August 12-14, 1996, Kansas City, Missouri.

National Research Council (1997), *Safe Water from Every Tap. Improving Water Service to Small Communities*, Committee on Small Water Supply Systems, Water Science and Technology Board, Commission on Geosciences, Environment, and Resources, National Research Council, National Academic Press, Washington, D.C.

NRC, see National Research Council.

Teuschler, J.M.; Goodrich, J.A.; Lykins Jr., B.W.; and Clark, R.M. (undated), *Remote Monitoring and Control of Small Drinking Water Package Plant Systems*, U.S. Environmental Protection Agency, Cincinnati, Ohio.

2

Small Treatment Systems Suitable for Remote Monitoring and Control

This chapter discusses the source water types, common violations, and applicable regulations (including promulgated and proposed maximum contaminant levels [MCLs] for all chemical and microbial contaminants regulated under the SDWAA) pertinent to small treatment systems and their operations. This chapter also assesses the amenability of typical small-system treatment technologies for remote monitoring and control. The amenability is evaluated based on the complexity and variability of the source water quality, as well as on process design and operation considerations. The monitoring and control needs of two example processes using groundwater as the source water are analyzed based on specific system operations and relevant water chemistry.

2.1 Current Status of Small Treatment Systems

Small treatment systems serving fewer than 10,000 people represent 94% of all water supply systems in the United States and serve 21% of the national population, according to a 1993 survey (MacDonald et al., 1997). Small treatment systems are commonly classified into three categories based on the population served: fewer than 500; 501 to 3,300; and 3,301 to 10,000 (U.S. EPA, 1989). The types of source water used by systems in each size category are summarized in Table 2-1. Groundwater is the most common source for small systems, and the use of groundwater is most predominant for the smallest system category.

TABLE 2-1. Source Water for Community Drinking Water Systems (NRC, 1997)

Population Served	Groundwater (%)	Surface Water (%)
<500	91	9
501-3,300	74	26
3,301-10,000	58	42
10,001-100,000	46	54
>100,001	28	72

Most small systems, primarily those in rural areas, do not have the resources to install control features in or to support qualified operators to operate their treatment trains. As a result, small systems are responsible for most violations of regulations under the SDWAA. The most common violation reported is for total coliform bacteria (NRC, 1997). A summary of systems with total coliform bacteria violations by size category from October 1, 1992 through December 31, 1994 is presented in Table 2-2. The second column in Table 2-2 shows the total number of systems for each size category in the United States, and the fourth column indicates the fraction of the total systems within each size category that reported violations. Almost 30% of the systems serving fewer than 500 people had total coliform bacteria violations, and most of these violations took place in systems using groundwater as the source water. Further, the percentage of violations corresponding to groundwater systems (fourth column in Table 2-2) was greater than the percentage of systems using groundwater for all system size categories (second column in Table 2-1).

TABLE 2-2. Total Coliform Bacteria Violations for the Period October 1, 1992 through December 31, 1994 (NRC, 1997)

	Systems with Violation		Violations by Source Water	
Population Served	Number of Systems	Percent of Total (%)	Groundwater Systems (%)	Surface Water Systems (%)
<500	10,509	29.5	95.0	5.0
501–3,300	1,938	13.4	84.8	15.2
3,301–10,000	592	14.4	71.8	28.2
>10,001	487	14.4	59.1	40.9

A summary of violations reported for chemical contaminants during the period of October 1, 1992 through December 31, 1994 is presented in Table 2-3. A comparison of the numbers presented in Tables 2-2 and 2-3 reveals that the number of chemical contaminant violations was much lower than the number of total coliform bacteria violations for all system size categories. For example, only 1.5% of all systems serving populations of fewer

TABLE 2-3. Chemical Contaminant Violations for the Period October 1, 1992 through December 31, 1994 (NRC, 1997)

	Systems with Violation		Violations by Source Water	
Population Served	Number of Systems	Percent of Total (%)	Groundwater Systems (%)	Surface Water Systems (%)
<500	531	1.5	96.4	3.6
501–3,300	162	1.1	73.5	26.5
3,301–10,000	25	0.6	60.0	40.0
>10,001	15	0.4	33.3	66.7

than 500 were in violation. Again, most of the violations took place in small systems using groundwater as the source water.

2.2 Applicable Drinking Water Regulations

Summaries of promulgated or proposed MCLs for all microbial and chemical contaminants regulated under the SDWAA are presented in Tables 2-4 through 2-6. Even though most of the regulated contaminants are chemicals, most violations are for microbial contaminants. This fact may change in the future as monitoring requirements change.

2.2.1 Disinfection Requirements for Surface Water

The microbial contaminant regulations summarized in Table 2-6 are those promulgated under the Surface Water Treatment Rule (SWTR) (U.S. EPA, 1989) or proposed in the Enhanced Surface Water Treatment Rule (ESWTR) (EPA, 1994b). The SWTR requires the practice of disinfection for all utilities treating surface water and groundwater under the influence of surface water. Most utilities also are required to filter their water unless the following conditions are met prior to disinfection:

- Fecal coliform bacteria <20/100 mL in 90% of samples
- Total coliform bacteria <100/100 mL in 90% of samples
- Turbidity <5 nephelometric turbidity units (NTUs)
- Other MCLs met.

Treatment plants exempted from filtration must disinfect to achieve 99.99% inactivation of viruses, and 99.9% inactivation of *Giardia lamblia* (*G. lamblia*) cysts. Compliance with these requirements must be demonstrated with the *CT* approach (the product of the average disinfectant concentration and t_{10} contact time). *CT* values estimated for actual disinfection systems must be equal to or greater than those summarized in Tables 2-7 and 2-8 for viruses and *G. lamblia* cysts, respectively. Treatment plants including some kind of filtration process

TABLE 2-4. Summary of Drinking Water Regulations (in μg/L) for Volatile Organic Compounds (VOCs) and Disinfection By-Products (DBPs) (U.S. EPA, 1991a, 1992, 1994a)

	Final Rule		Proposed Rule		
Contaminant	**MCL**	**MCLG**	**MCL**	**MCLG**	**Date**
VOCs					
Benzene	5	0			1/91
Carbon Tetrachloride	5	0			1/91
1,2-Dibromo-3-Chloropropane	0.2	0			1/91
o-Dichlorobenzene	600	600			1/91
p-Dichlorobenzene	75	75			1/91
1,2-Dichloroethane	5	0			1/91
1,1-Dichloroethylene	7	7			1/91
cis-1,2-Dichloroethylene	70	70			1/91
trans-1,2-Dichloroethylene	100	100			1/91
1,2-Dichloropropane	5	0			1/91
Ethylbenzene	700	700			1/91
Ethylene Dibromide	0.05	0			1/91
Methylene Chloride	5	0			7/92
Monochlorobenzene	100	100			1/91
Tetrachloroethylene	5	0			1/91
Toluene	1,000	1,000			1/91
1,2,4-Trichlorobenzene	70	70			7/92
1,1,1-Trichloroethane	200	200			1/91
1,1,2-Trichloroethane	5	3			7/92
Trichloroethylene	5	0			1/91
Styrene	100	100			1/91
Vinyl Chloride	2	0			1/91
o-, *m*-, and *p*-Xylenes	10,000	10,000			1/91
Organic DBPs					
Trihalomethanes (THMs)					
Bromodichloromethane				0	7/94
Bromoform				0	7/94
Chloroform				0	7/94
Dibromochloromethane				60	7/94
Total THMs			80⇒40		7/94
Haloacetic Acids (HAAs)					
Dichloroacetic Acid				0	7/94
Trichloroacetic Acid				300	7/94
Total HAAs			60⇒30		7/94
Chloral Hydrate			—	40	
Inorganic DBPs					
Bromate			10	0	7/94
Chlorite			1,000	80	7/94

Note: MCLG is maximum contaminant level goal.

TABLE 2-5. Summary of Drinking Water Regulations for Synthetic Organic Compounds (SOCs) (in μg/L) (U.S. EPA, 1991a,c; 1992)

Contaminant	Final Rule MCL	Final Rule MCLG	Proposed Rule MCL	Proposed Rule MCLG	Date
Other SOCs					
Acrylamide	T.T.[a]	0			1/91
Adipates					
Di(2-ethylhexyl) adipate	400	400			7/92
Alachlor	2	0			1/91
Aldicarb	3	1			7/91
Aldicarb Sulfone	3	1			7/91
Aldicarb Sulfoxide	4	1			7/91
Atrazine	3	3			1/91
Carbofuran	40	40			1/91
Chlordane	2	0			1/91
2,4-D	70	70			1/91
Dalapon	200	200			7/92
Dinoseb	7	7			7/92
Diquat	20	20			7/92
Endothall	100	100			7/92
Endrin	2	2			7/92
Epichlorohydrin	T.T.	0			1/91
Glyphosate	700	700			7/92
Heptachlor	0.4	0			1/91
Heptachlor Epoxide	0.2	0			1/91
Hexachlorobenzene	1	0			7/92
Hexachlorocyclopentadiene	50	50			7/92
Lindane	0.2	0.2			1/91
Methoxychlor	40	40			1/91
Pentachlorophenol	1	0			7/91
Phthalates					
Di(2-ethylhexyl) phthalate	6	0			7/92
Pichloram	500	500			7/92
Polychlorinated Biphenyls (PCBs) (as decachlorobiphenyl)	0.5	0			1/91
Polycyclic Aromatic Hydrocarbons (PAHs)					
Benzo(a)pyrene	0.2	0			7/92
Benz(a)anthracene			0.1	0	7/90
Benzo(b)fluoranthene			0.2	0	7/90
Benzo(k)fluoranthene			0.2	0	7/90
Chrysene			0.2	0	7/90
Dibenz(a,h)anthracene			0.3	0	7/90
Indenopyrene			0.4	0	7/90
Simazine	4	4			7/92
2,3,7,8-Tetrachlorodibenzodioxin (Dioxin)	0.0000	0			7/92
Toxaphene	3	0			1/91
2,4,5-Trichlorophenoxypropionic Acid (Silvex)	50	50			1/91
Vydate (Oxamyl)	200	200			7/92

(a) T.T. = treatment technique.

TABLE 2-6. Summary of Drinking Water Regulations (in μg/L unless otherwise indicated) for Inorganic Compounds (IOCs) and Microbial Contaminants (U.S. EPA, 1975; 1989; 1991a,b,c,d; 1992; 1994a,b,c)

	Final Rule		Proposed Rule		
Contaminant	**MCL**	**MCLG**	**MCL**	**MCLG**	**Date**
IOCs					
Antimony	6	6			7/92
Arsenic	50	0			12/75
Asbestos	7 MF/L[(a)]	7 MF/L[(a)]			1/91
Barium	2,000	2,000			7/91
Beryllium	4	4			7/92
Cadmium	5	5			1/91
Chromium	100	100			1/91
Copper	T.T.[(b)]	1,300			6/91
Cyanide (as free Cyanide)	200	200			7/92
Fluoride	4,000	4,000			7/91
Lead	T.T.	0			6/91
Mercury	2	2			1/91
Nickel	100	100			7/92
Nitrate + Nitrite (as N)	10,000	10,000			1/91
Nitrite (as N)	1,000	1,000			1/91
Selenium	50	50			1/91
Sulfate			500,000	500,000	12/94
Thallium	2	0.5			7/92
Disinfectants					
Chloramines (as Cl_2)			4,000[(d)]	4,000[(d)]	7/94
Chlorine (as Cl_2)			4,000[(d)]	4,000[(d)]	7/94
Chlorine Dioxide (as ClO_2)			800[(d)]	300[(d)]	7/94
Radionuclides					
Beta and Photon Emitters			4[(e)]	0[(e)]	7/91
Gross Alpha Emitters			15 pCi/L	0	7/91
Radium-226 and 228			20 pCi/L	0	7/91
Radon-222			300 pCi/L	0	7/91
Uranium			20[(f)]	0	7/91
Microbial Contaminants					
Cryptosporidium parvum Oocysts			T.T.	0	7/94
Giardia lamblia Cysts			T.T.	0	7/94
Viruses			T.T.	0	7/94
Legionella	T.T.	0			6/89
Heterotrophic Bacteria	T.T.	—			6/89
Total Coliforms	Absence[(c)]	0			6/89
Turbidity	T.T.	—			6/89

(a) Asbestos MCL and MCLG values expressed in million fibers (longer than 10 μm) per liter.
(b) T.T. = treatment technique.
(c) Fewer than 1 positive sample/month (small utility), or fewer than 5% positive samples (large utility).
(d) Proposed disinfectant standards for chloramines, chlorine, and chlorine dioxide are defined in terms of maximum residual disinfectant levels (MRDLs) and MRDL goals.
(e) MCL and MCLG values of beta particles and photon emitters expressed in mrem/yr (dose committed over a period of 50 years to reference man from an annual intake at the rate of 2 L of drinking water per day).
(f) 20 μg/L of uranium correspond to 30 pCi/L.

are credited with a certain level of microorganism removal, with the balance to be treated by disinfection. For example, conventional filtration (i.e., coagulation, flocculation, sedimentation, and filtration) plants are credited with 2.0 and 2.5 log (i.e., 99.0 and 99.7%) removal of viruses and *G. lamblia* cysts, respectively. Consequently, the disinfection process must be designed to treat at least an additional 2.0 and 0.5 log (i.e., 99.0 and 68.4%) of viruses and *G. lamblia* cysts, respectively. Alternative filtration processes including direct, slow sand, and diatomaceous earth filtration are each credited with 2.0 log removal of *G. lamblia* cysts (i.e., control of the remaining 1.0 log must be accomplished by disinfection). Direct and diatomaceous earth filtration processes are each credited with 1.0 log removal, and slow sand filtration with 2.0 log removal for viruses. The balance to the overall 4.0 log treatment must be accomplished by disinfection. Minimum *CT* requirements to achieve the levels of various minimum levels of inactivation with common drinking water disinfectants (free chlorine, ozone, chlorine dioxide, and chloramines) are summarized in Tables 2-7 and 2-8.

TABLE 2-7. *CT* Values (mg × min/L) for Virus Inactivation (U.S. EPA, 1989)

Disinfectant	Log Inactivation	Temperature 0.5°C	5°C	10°C	15°C	20°C	25°C
Free Chlorine	2.0	6	4	3	2	1	1
(pH = 6-9)	3.0	9	6	4	3	2	1
	4.0	12	8	6	4	3	2
Ozone	2.0	0.9	0.6	0.5	0.3	0.25	0.15
(pH = 6-9)	3.0	1.4	0.9	0.8	0.5	0.4	0.25
	4.0	1.8	1.2	1.0	0.6	0.5	0.3
Chlorine Dioxide	2.0	8.4	5.6	4.2	2.8	2.1	1.4
(pH = 6-9)	3.0	25.6	17.1	12.8	8.6	6.4	4.3
	4.0	50.3	33.5	25.1	16.8	12.6	8.4
Chloramines	2.0	1,243	857	643	428	321	214
(pH = 6-9)	3.0	2,063	1,423	1,067	712	534	356
	4.0	2,883	1,988	1,491	994	746	497

2.2.2 Disinfection Requirements for Groundwater

The control of microbial contaminants in drinking water systems using groundwater as the source water is regulated primarily at state level and will be regulated under the Groundwater Disinfection Rule (GWDR) currently being developed. The GWDR most likely will address the following issues (Macler, 1997):

- Protection of groundwater source
- Integrity of well and distribution system
- Protection against distribution system failure
- Disinfection
- Monitoring of wellhead and system for fecal contamination.

TABLE 2-8. ***CT*** **Values (mg × min/L) for** ***Giardia lamblia*** **Oocyst Inactivation (U.S. EPA, 1989)**

Disinfectant	Log Inactivation	Temperature					
		0.5°C	5°C	10°C	15°C	20°C	25°C
Free Chlorine	0.5	25	18	13	9	7	4
(pH = 6)	1.0	49	35	26	18	13	9
	3.0	165	116	87	58	44	29
Free Chlorine	0.5	35	25	19	13	9	6
(pH = 7)	1.0	70	50	37	25	19	12
	3.0	236	165	124	83	62	41
Free Chlorine	0.5	51	36	27	18	14	9
(pH = 8)	1.0	101	72	54	36	27	18
	3.0	346	243	182	122	91	61
Free Chlorine	0.5	73	52	39	26	20	13
(pH = 9)	1.0	146	104	78	52	39	26
	3.0	500	353	265	177	132	88
Ozone	0.5	0.48	0.32	0.23	0.16	0.12	0.08
(pH = 6-9)	1.0	0.97	0.63	0.48	0.32	0.24	0.16
	3.0	2.9	1.9	1.4	0.95	0.72	0.50
Chlorine Dioxide	0.5	10	7	5	3.3	3	1.7
(pH = 6-9)	1.0	20	13	10	5	5	3.3
	3.0	60	40	30	20	15	10
Chloramines	0.5	690	363	337	250	181	130
(pH = 6-9)	1.0	1,295	737	675	505	366	260
	3.0	3,800	2,200	1,850	1,500	1,100	750

The current trend of thought in the GWDR is that disinfection should be practiced to provide an additional barrier to protect against fecal contamination (Macler and Pontius, 1997).

2.3 Treatment Technologies Used by Small Systems

A summary of treatment technologies currently used by small water treatment systems is presented in Table 2-9. Many of these processes are available as package plants (Campbell et al., 1995; Clark et al., 1994; Goodrich et al., 1992), with some degree of automation designed specifically for small systems. Processes currently in most common use by small systems include disinfection for both groundwater and surface water, and the various filter configurations having impact on small systems including membrane filtration processes, i.e., microfiltration (MF) and ultrafiltration (UF), for the treatment of surface water, and nanofiltration (NF) and reverse osmosis (RO) for the control of groundwater contaminants.

Small treatment systems can be grouped into three categories: full-scale treatment plants, centrally located package plants, and point-of-entry/point-of-use (POE/POU) devices. These systems are discussed in Sections 2.3.1 through 2.3.3.

TABLE 2-9. Treatment Technologies Commonly Used for Small Water Systems (NRC, 1997)

Source Water	Technology
Groundwater	Air Stripping Oxidation/Filtration Ion Exchange Activated Alumina
Surface Water	Coagulation/Filtration Dissolved Air Flotation Diatomaceous Earth Filtration Slow Sand Filtration Bag and Cartridge Filtration
Groundwater and Surface Water	Disinfection Corrosion Control Membrane Filtration Systems Reverse Osmosis/Nanofiltration Electrodialysis Systems Adsorption Lime Softening

2.3.1 Full-Scale Treatment Plants

Full-scale treatment plants may employ filtration (including slow sand, diatomaceous earth, and reverse osmosis [RO] membranes) and disinfection (including the use of chlorine, ozone, chlorine dioxide, or chloramines) to remove a variety of substances, such as particulate, microorganisms, color, and disinfection by-product (DBP) precursors. These plants also may use processes, such as granular activated carbon (GAC) adsorption, packed-tower or diffused aeration, advanced oxidation, and RO to control organic contaminants, and processes such as RO, ion exchange (IX), GAC, and activated alumina (AA) to control inorganic contaminants.

Figures 2-1 and 2-2 show two treatment processes of a full-scale treatment plant that treats groundwater with aeration, sand filtration, RO, and postchlorination for 2,300 households with about 7,300 people. The two-train RO system is used to remove hardness and mineral content (from 650 and 450 mg/L to 150 and 120 mg/L, respectively, after blending).

2.3.2 Centrally Located Package Plants

Many processes can be incorporated into a package plant. The technologies most commonly used include filtration, membranes, GAC, and aeration. Disinfection often is used to provide a disinfectant residual, especially for the systems using surface water as the source water, per regulations promulgated under the SWTR or proposed in the ESWTR. The control of microbial contaminants for systems using groundwater as the source water is regulated at state level and will be regulated under the GWDR currently being developed.

FIGURE 2-1. Sand filter in a full-scale small treatment plant in Central Ohio (filter was being backwashed when photograph was taken).

FIGURE 2-2. Reverse osmosis system in a full-scale small treatment plant in Central Ohio.

2.3.2.1 Filtration

More than 1,000 filtration package plants operate in the United States. Many of these plants are built to conventional design standards. The three basic types are conventional package plants, tube-type clarification package plants, and adsorption clarifier package plants. The conventional plants (see Figure 2-3) usually contain conventional processes such as coagulation, flocculation, sedimentation, and filtration. The tube-type clarifier plants (see Figure 2-4) include the use of settling tubes to increase the clarification efficiency and reduce the system size. The adsorption clarifier plants (see Figure 2-5) use adsorptive clarifiers consisting of an upflow filter with low-density plastic bead media to replace the flocculation and sedimentation basins.

FIGURE 2-3. A conventional package plant consisting of coagulation, flocculation, sedimentation, and filtration.

Evaluations of 19 municipal and 17 recreational filtration package plants indicate that the filtration systems can meet the traditional goals of controlling turbidity and microbiological contaminants (Clark et al., 1994). Although not tested, the removal of *G. lamblia* cysts and enteric viruses is not expected to be satisfactory, because these systems are designed primarily for turbidity control. Difficulties in meeting the MCLs usually are caused by lack of operator attention, such as not using the proper chemical dosage for changing raw water quality or not running the systems long enough to achieve stable operations.

FIGURE 2-4. Pulsapak® package water treatment system using a packed sludge bed that incorporates settling tubes. (Photograph courtesy of Infilco Degremont, Inc. Used with permission.)

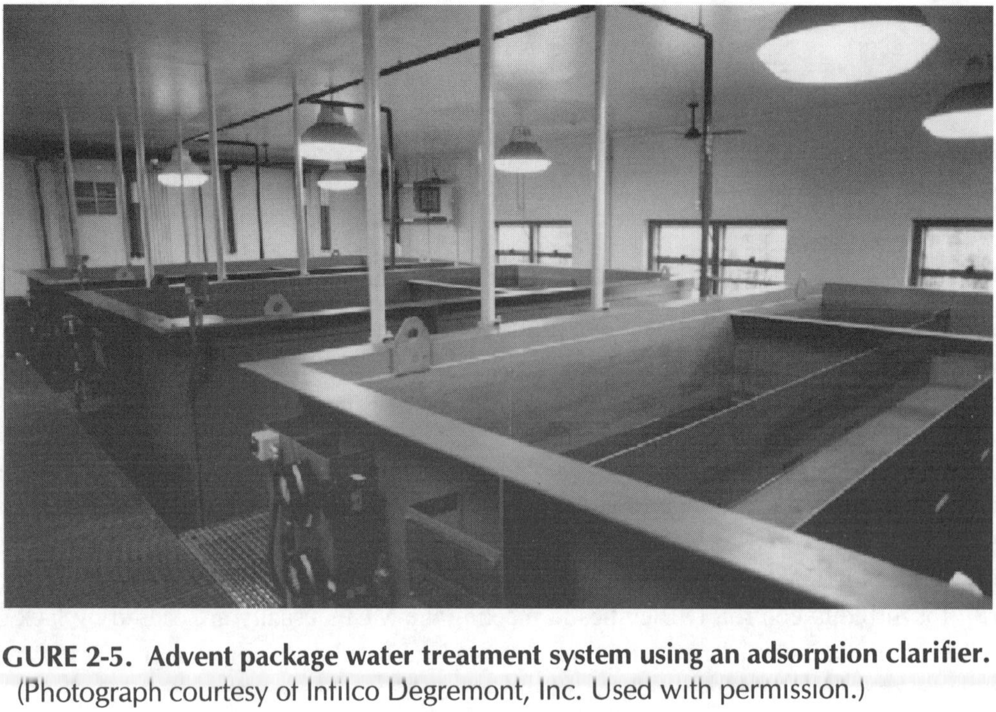

FIGURE 2-5. Advent package water treatment system using an adsorption clarifier. (Photograph courtesy of Infilco Degremont, Inc. Used with permission.)

2.3.2.2 Membranes

Based on nominal pore size or molecular weight cutoff (MWCO), membrane filtration processes may be classified into four major categories: MF, UF, NF, and RO (see Figure 2-6 for sizes of molecules/particles/microbes removed by various membrane pore sizes). MF, UF, and NF processes use membranes arranged in either hollow-fiber or spiral-wound configuration. A hollow-fiber module has a very large membrane area per unit volume and a relatively low pressure drop within the module when compared to a spiral-wound module. However, the hollow-fiber module may have a channeling problem when operated in a shell-feed mode, and a fouling problem if the feed contains large amounts of particulate or other foulants. Figure 2-7 presents a UF system with a fully automated control panel. Also shown in the figure are the schematics of hollow-fiber and spiral-wound membrane modules.

NF and RO filtration can effectively remove protozoa, bacteria, and even viruses, thus relieving a portion of the *CT* disinfection requirement during the treatment. The presence of other particulate matter, however, may cause premature membrane fouling. Therefore, surface water with high turbidity and solids content may require pretreatment by a more conventional treatment process prior to membrane filtration. For groundwater sources not requiring

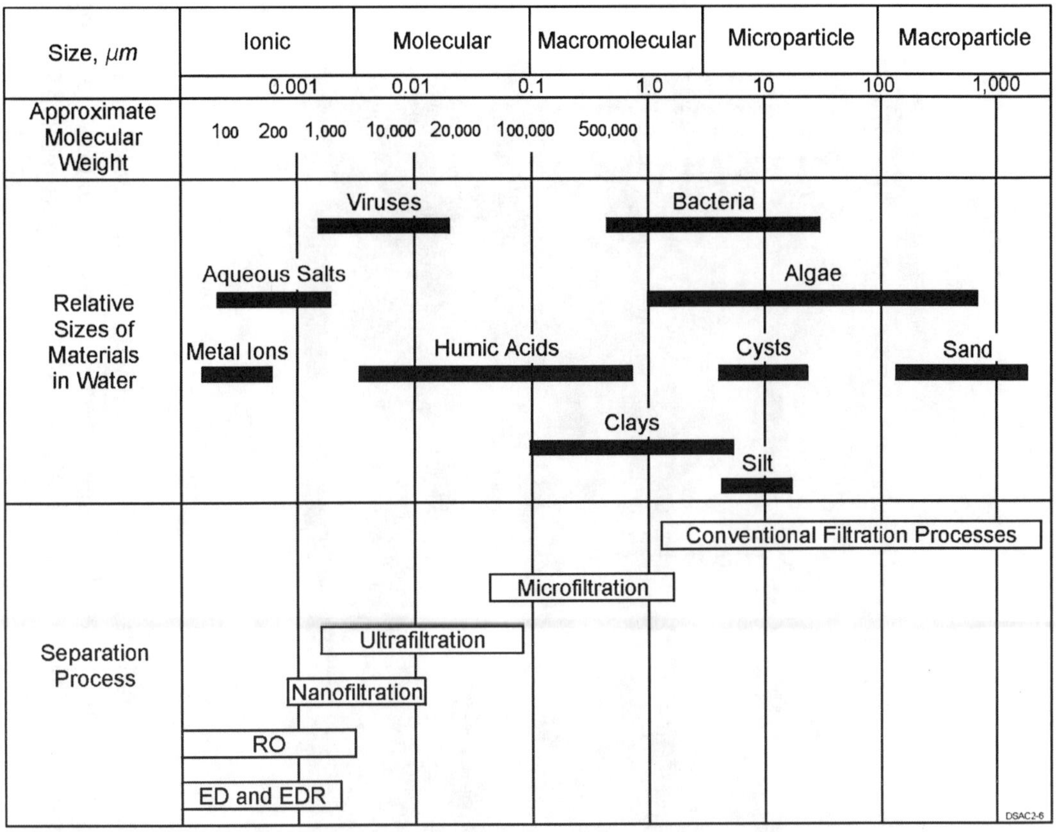

FIGURE 2-6. Sizes of molecules removed by various membrane processes in comparison to conventional filtration processes (After AWWA, 1995).
ED = electrodialysis; EDR = electrodialysis reversal.

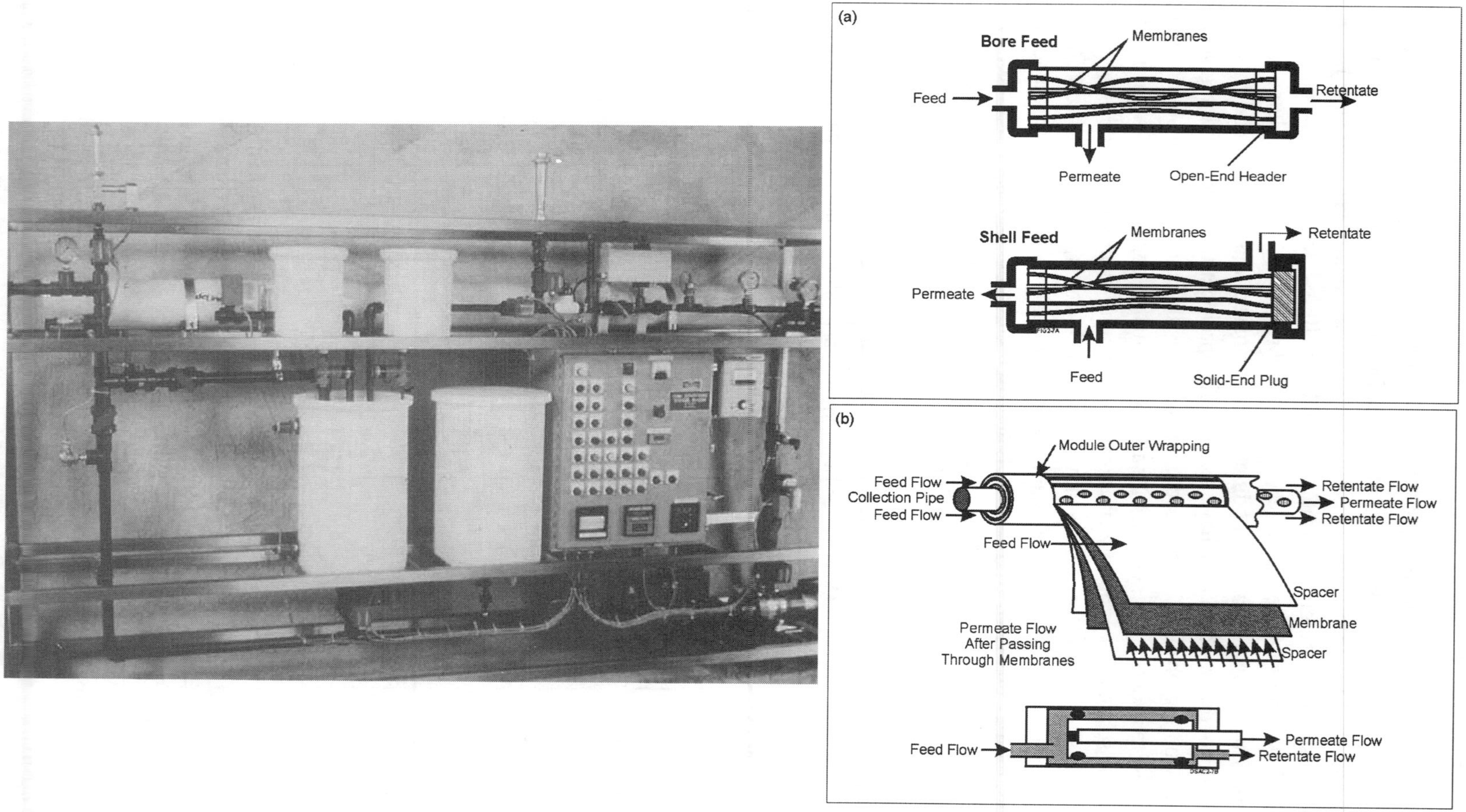

FIGURE 2-7. An ultrafiltration package plant with (a) a hollow-fiber and (b) a spiral-wound membrane module. Insets (a) and (b) are adopted from *Chemical Engineering Progress 91*(10):68, 1995. Reproduced with permission of the American Institute of Chemical Engineers.

conventional pretreatment, cartridge/bag filters often are used to remove particles and particulate matter. Figure 2-8 presents typical bag filter cartridges operating in series. When bag filters are the only required pretreatment, membrane filtration becomes a very attractive option because it is easy to install and operate.

2.3.2.3 Granular Activated Carbon Adsorption

GAC adsorption can be easily applied to small treatment systems. It is used primarily to control tastes and odors and, more recently, toxic or carcinogenic organic compounds. GAC may be used to perform filtration and adsorption in a single step, which would require more frequent backwashing. Backwashing mixes the GAC in the column, causing some spent carbon to be placed near the effluent. The effluent stream, therefore, may contain some of the target compounds. Other concerns over the use of GAC are bacterial regrowth and competitive adsorption between the target compounds and naturally occurring organic matter. The design of GAC plants must consider these possibilities. Figure 2-9 shows an online GAC system that uses RO and a tray aerator.

2.3.2.4 Aeration

Aeration often is used in conjunction with filtration for groundwater treatment. Aeration involves continuous contact of air with water to strip volatile organic compounds (VOCs),

FIGURE 2-8. Typical bag filter cartridges operating in series.

FIGURE 2-9. A GAC and an RO system on line with a tray aerator (shown in Figure 2-10).

taste- and odor-causing compounds, and perhaps radon gas, and to oxidize iron and manganese. The stripped air is treated, if necessary, and released to the atmosphere, and the precipitated iron and manganese are removed by filtration. Aeration usually is accomplished through the use of compressed air and a series of diffusers (i.e., diffused aeration); agitation with a mechanical mixer (i.e., mechanical aeration); a vertical series of trays (i.e., tray aerators, Figure 2-10); or a bed of packing material, such as fixed plastic grids, loose plastic rings, or loose ceramic saddles (i.e., packed-tower aerators, Figure 2-11). Recently, the use of gas-permeable membrane modules made up of highly porous, hollow-fiber membranes, has been gaining acceptance. Because of the large available surface area, this technology offers the highest removal efficiency of all aerators. The gas-permeable membrane system, however, has not been fully tested.

2.3.3 Point-of-Entry/Point-of-Use Devices

POE treatment units may be used by industries, apartment buildings, restaurants, trailer parks, and homes. POU units may be used at a single faucet. POE/POU devices are becoming attractive alternatives for small communities because they are readily available, relatively inexpensive, and easy to install and maintain. Currently, the U.S. EPA accepts POE treatment as a technology for complying with MCLs, but does not accept POU devices.

FIGURE 2-10. A tray aerator with a storage reservoir shown in the background and a secured building in the foreground.

The public water systems equipped with POE treatment must adhere to a set of rules for compliance. Figure 2-12 shows a multimedia/O_3/ultraviolet (UV)/GAC POE system with the O_3/UV closeup shown on the right.

2.4 Amenability of Treatment Technologies

Table 2-10 presents the amenability of various treatment technologies to remote monitoring and control. The amenability can be assessed based on the complexity and variability of the source water quality, as well as process design and operational considerations. In general, it might be possible to find site-specific conditions that could result in successful plant automation and remote control for most of these processes. For example, air-stripping

FIGURE 2-11. A packed-tower aerator packed with loose plastic grids.

towers commonly used to treat groundwater contaminated with VOCs could be automated and remote-controlled when treating groundwater of acceptable quality other than containing refractory VOCs. In contrast, process automation might not be viable if the groundwater also is contaminated with solutes such as reduced iron or aerobically biodegradable organic matter, conditions commonly resulting in quick fouling/biofouling and the corresponding mass transfer deterioration and overall process failure.

Chlorine and potassium permanganate are commonly used to oxidize reduced iron and manganese in small systems treating groundwater. The resulting precipitates can be removed by filtration. However, these processes are not amenable to automation and remote control due to the complexity of the chemistry involved.

Small, automated ion-exchange units are commercially available for removal of cations (such as hardness) and anions (such as arsenate and arsenite ions). Although ion-exchange technology can be used as part of small treatment systems, its application is more common as a POE or POU device.

Activated alumina can be used to remove fluoride and arsenic from the water supplied to small communities. However, its application is more common as a POE or POU device, and automation/remote control of this process is limited by its chemical complexity.

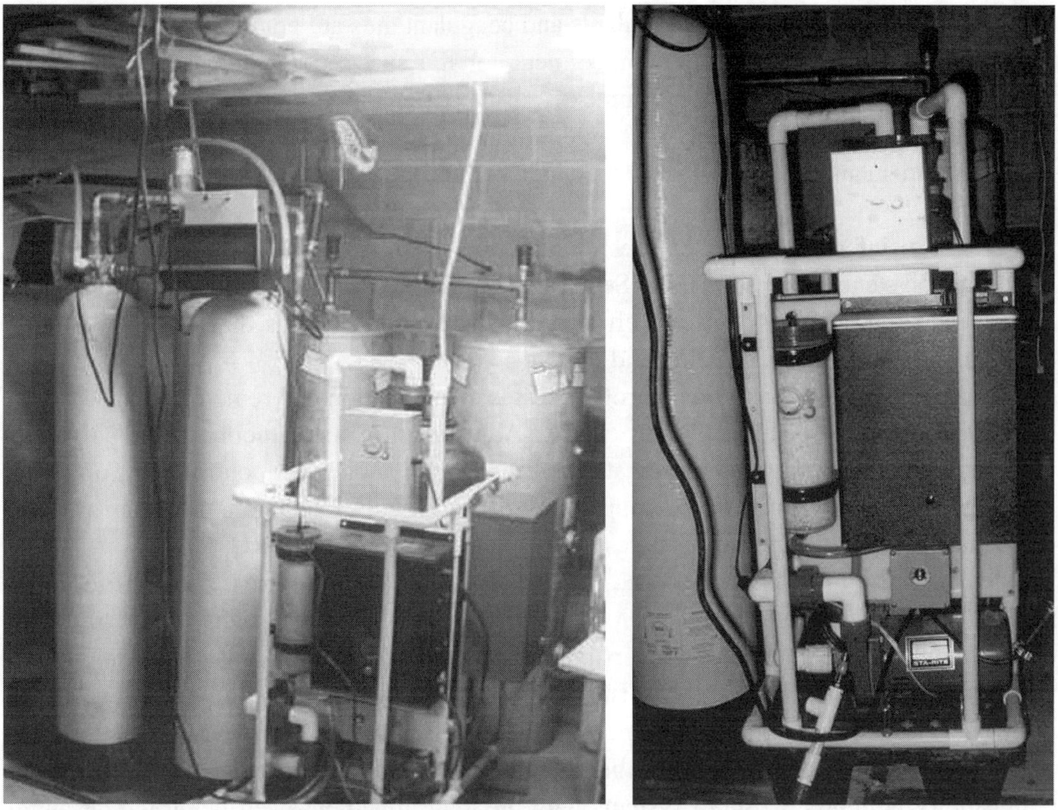

FIGURE 2-12. A multimedia/O_3/UV/GAC POE system. A closeup of the O_3/UV component is shown on the right.

TABLE 2-10. Amenability of Treatment Technologies Used for Small Water Systems to Automation and Remote Control (NRC, 1997)

Source Water	Technology	Amenability for Automation/Remote Control
Groundwater	Air Stripping	Very Good to Poor
	Oxidation/Filtration	Poor
	Ion Exchange	Good to Poor
	Activated Alumina	Poor
Surface Water	Coagulation/Filtration	Poor
	Dissolved Air Flotation	Poor
	Diatomaceous Earth Filtration	Good to Poor
	Slow Sand Filtration	Good to Poor
	Bag and Cartridge Filtration	Good to Poor
Groundwater and Surface Water	Disinfection	Very Good
	Corrosion Control	Good to Poor
	Membrane Filtration Systems	Good to Poor
	Reverse Osmosis/Nanofiltration	Very Good to Poor
	Electrodialysis Systems	Very Good
	Adsorption	Good to Poor
	Lime Softening	Poor

Filtration processes that use coagulants and coagulant aids are common in small systems that treat surface water. A major impediment for the successful automation and remote control of these processes is the complexity of the coagulation chemistry and the lack of consistency between the coagulant dosage requirement and degree of water quality. Thus, filtration and membrane systems that do not include coagulation are more amenable to automation.

Among disinfection processes, free chlorination is the simplest and most adaptable for most small systems. As discussed in Section 2.2.1, the *CT* approach can be used to demonstrate adequate system disinfection efficiency. Disinfection systems that use free chlorine are highly amenable to automation and remote control.

The amenability of corrosion control processes to automation and remote control depends on the type of process and the water quality. Adjustment of the orthophosphate/polyphosphate dosage (1 to 3 mg/L as phosphate) and the corresponding pH control (e.g., optimum pH range of 7.2 to 7.8 for lead and copper control) is difficult. Excursions from narrow optimum ranges might induce various types of water quality problems (NRC, 1997).

Membrane filtration processes (i.e., MF, UF, NF, and RO have great potential for automation and remote control. However, variations in surface water quality could result in exceedingly fast and sometimes irreversible membrane fouling. In contrast, because groundwater quality tends to be more consistent, NF and RO systems treating groundwater are more amenable to automation and remote control as discussed in Section 2.5.2.

Electrodialysis systems are amenable to automation and remote control. However, their applicability is limited to small systems with the main treatment need being the desalination of brackish water.

Powdered activated carbon (PAC) and GAC processes are suitable for small systems requiring removal of organic contaminants. However, the amenability of these processes to automation and remote control is limited by the complexity of the analytical systems required to monitor organic contaminant(s) breakthrough curves.

Lime softening can be used in small treatment systems to remove hardness. Its amenability to automation and remote control, however, is limited by the chemical complexity of the process.

2.5 Two Example Processes

Based on the discussion in Section 2.4 and the current status of small systems with respect to regulatory compliance (see Tables 2-2 and 2-3), disinfection and NF are two example processes amenable to remote monitoring and control. Therefore, they are used to illustrate the RTS concept. Because groundwater has been the main source used by most small systems, it is targeted for treatment by these processes. The use of groundwater as the main source is consistent with the recommendation of the NRC urging small systems to make a strong effort to tap into groundwater sources (NRC, 1997). The monitoring and control needs of the disinfection and NF processes are discussed in Sections 2.5.1 and 2.5.2.

2.5.1 Groundwater Disinfection

A schematic of a typical groundwater disinfection plant is shown in Figure 2-13. Plant components include a well and the associated pump, chemical addition, contacting chamber/clear well, and disinfectant residual monitoring and control system. Chemicals added include the disinfectant (most commonly free chlorine) and, in some cases, conditioning compounds such as bases or other corrosion inhibitors. The plant can be automated easily by measuring the free chlorine residual, pH, temperature, and flowrate of the water exiting from the contacting chamber or clear well. The results from tracer test studies may be used to develop a correlation between the flowrate and regulatory contact time, t_{10}. The control system would then include software to compute the CT value by multiplying the residual concentration (C) and the corresponding t_{10} contact time (T). The resulting CT may be compared to minimum required values at the actual temperature and pH of the specific water. The minimum CT values can be estimated by linear interpolation of the values presented in Tables 2-7 and 2-8. The control system should be capable of increasing or decreasing the disinfectant feeding pump as needed to achieve a target CT value somewhat greater than the minimum. The target value may be set based on operating experience with specific systems.

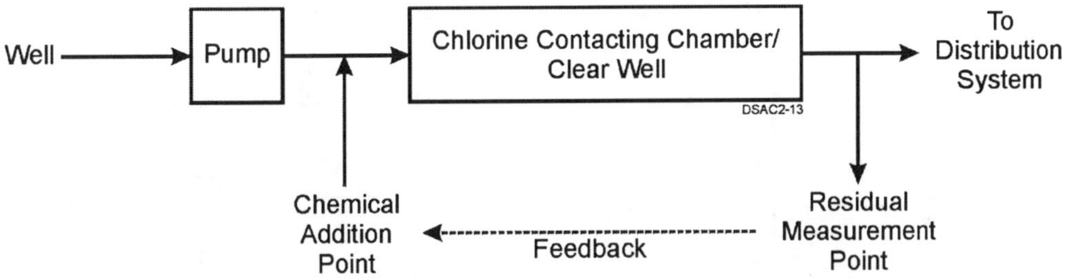

FIGURE 2-13. Schematic of disinfection plant using free chlorine for groundwater treatment.

An important consideration with groundwater is the presence of ammonia nitrogen and its variable concentration. Chlorine will tend to react with ammonia, forming chloramines at relatively low chlorine:ammonia ratios. Because chloramines are much weaker disinfectants compared to free chlorine, the dosage of chlorine must be increased to levels at which ammonia can be oxidized (i.e., breakpoint chlorination). The use of a probe capable of detecting only free chlorine could allow the control system to dose chlorine levels resulting in breakpoint chlorination. The chlorine dose required will be that necessary to oxidize ammonia (typically 1.5 to 2.0 times the concentration of ammonia on a molar basis) plus the residual corresponding to the target CT value.

2.5.2 Nanofiltration

A schematic of an NF plant for the treatment of groundwater is presented in Figure 2-14. The same schematic will apply to the case of an RO plant. The system includes a well

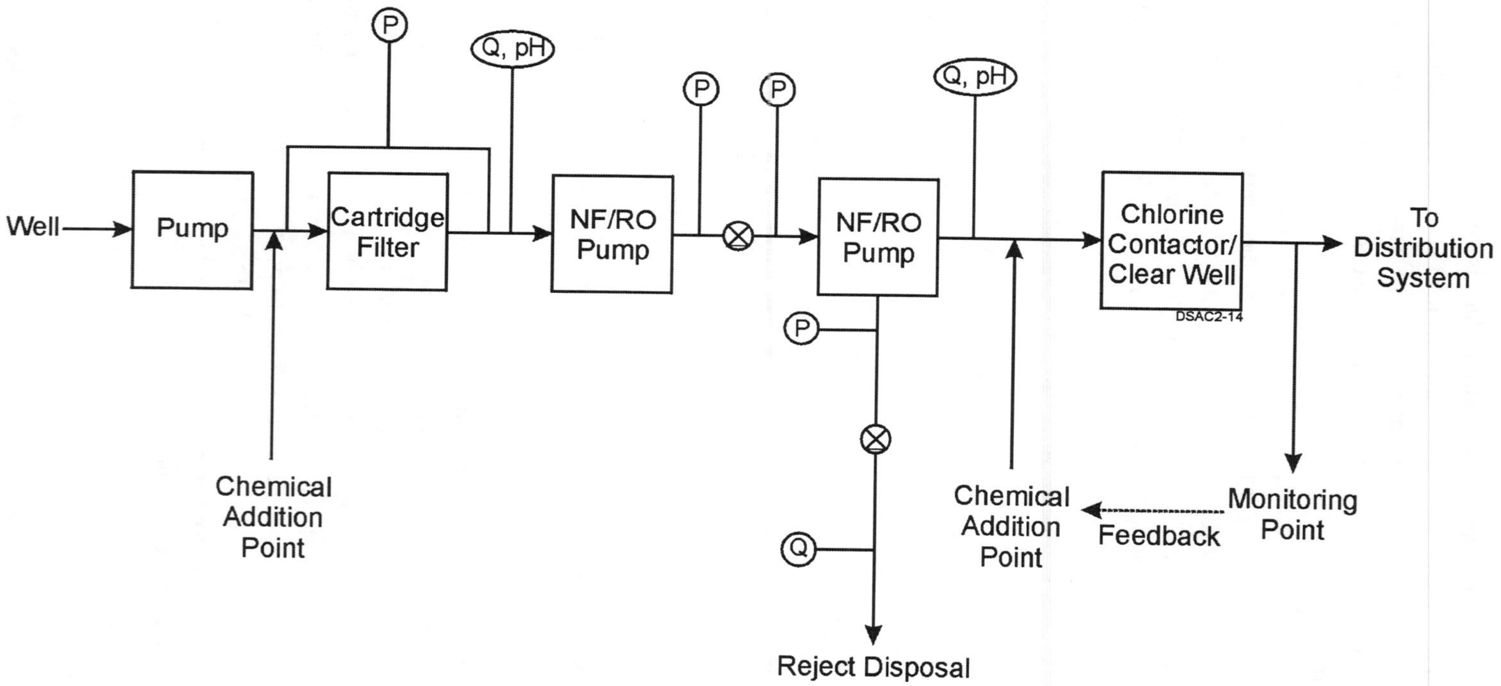

FIGURE 2-14. Schematic of nanofiltration plant for groundwater treatment, where P is a pressure monitoring location and Q is where the quantity of water is monitored.

and the corresponding pump, chemical pretreatment system, cartridge filter, NF pump, NF membrane system, chemical stabilization and disinfectant addition system, contact chamber or clear well, and disinfectant residual monitoring and control system.

Chemical pretreatment typically consists of the addition of a concentrated acid and/or a scale inhibitor to prevent the precipitation of sparingly soluble salts (e.g., calcium carbonate, calcium sulfate, calcium fluoride, etc.) as the reject water is concentrated in the NF membrane system. The target pH value can be between 4.5 and the natural pH of the water, depending on the precipitation potential of the specific water of interest. The concentrated acid dose requirement will be affected by the alkalinity of the water. Scale inhibitors tend to be chemicals of proprietary composition. The manufacturer of each commercial product typically provides design dosage recommendations based on water quality information.

The typical configuration of the cartridge filter is a vessel encasing one or more cartridges with the mean pore size diameter in the range of 0.5 to 5 μm. The unit is operated continuously until the differential pressure exceeds 10 to 15 pounds per square inch (psi), at which time the cartridge(s) must be replaced.

The NF pump is used to pressurize the water prior to entering the NF membrane system. The feed flow to the membrane system must remain approximately constant. Only a fraction of the flow fed to the membrane system, most commonly 75 to 85%, is treated to become product water. The residual or reject water is depressurized and disposed of in accordance with a permit negotiated with the regulatory agency. The NF membrane system typically is operated at a constant product-water flowrate. However, membranes tend to undergo fouling and, as a result, the product-water flowrate tends to decrease. In the case of a constant-speed NF pump, this problem can be overcome by varying the opening setting of the valve located between the NF pump and the membrane system. A second approach that could result in a constant flowrate while varying the pressure is the use of variable-speed centrifugal pumps.

An automated disinfection system similar to that presented in Figure 2-13 could also be used to treat the product water from the membrane system. If required, the water may be stabilized by adjusting the pH value.

2.6 References

American Water Works Association (1995), *Electrodialysis and Electrodialysis Reversal.* Denver, Colorado.

AWWA, see American Water Works Association

Campbell, S.; Lykins Jr., B.W.; Goodrich, J.A.; Post, D.; Lay, T. (1995), Package Plants for Small Systems: A Field Study, *Journal of the American Water Works Association, 95(11),* pp. 39-47.

Clark, R.M.; Goodrich, J.A.; Lykins Jr., B.W. (1994), Package Plants for Small Water Supplies—The U.S. Experience, *Journal Water SRT, 43(1),* pp. 23-34.

Goodrich, J.A.; Adams, J.Q.; Lykins Jr., B.W.; Clark, R.M. (1992), Safe Drinking Water from Small Systems: Treatment Options, *Journal of the American Water Works Association, 92(5),* pp. 49-55.

MacDonald, J.A.; Zander, A.K.; Snoeyink, V.L. (1997), Improving Service to Small Communities, *Journal of the American Water Works Association, 89(1),* pp. 58-64.

Macler, B.A. (1997), Scientific Issues for the Groundwater Disinfection Rule, *Proceedings of the American Water Works Association Water Quality Technology Conference,* November 9-12, 1997, Denver, Colorado.

Macler, B.A.; Pontius, F.W. (1997), Update on the Ground Water Disinfection Rule, *Journal of the American Water Works Association, 89(1)*, pp. 16-20 and 115.

National Research Council (1997), *Safe Water from Every Tap. Improving Water Service to Small Communities*, Committee on Small Water Supply Systems, Water Science and Technology Board, Commission on Geosciences, Environment, and Resources, National Research Council, National Academic Press, Washington, D.C.

NRC, see National Research Council.

U.S. Environmental Protection Agency (1975), National Interim Primary Drinking Water Regulations, *Federal Register, 40(248)*, 59566-59588, December 24, 1975.

U.S. Environmental Protection Agency (1989), 40 CFR Parts 141 and 142; Drinking Water; National Primary Drinking Water Regulations; Filtration, Disinfection; Turbidity, *Giardia lamblia*, Viruses, Legionella, and Heterotrophic Bacteria; Final Rule, *Federal Register, 54(124)*, 27486-27541, June 19, 1989.

U.S. Environmental Protection Agency (1991a), 40 CFR Parts 141, 142, and 143; National Primary Drinking Water Regulations; Final Rule, *Federal Register, 56(20)*, 3526-3597, January 30, 1991.

U.S. Environmental Protection Agency (1991b), 40 CFR Parts 141 and 142; Maximum Contaminant Level Goals and National Primary Drinking Water Regulations for Lead and Copper; Final Rule, *Federal Register, 56(110)*, 26460-26564, June 7, 1991.

U.S. Environmental Protection Agency (1991c), 40 CFR Parts 141, 142, and 143; National Primary Drinking Water Regulations; Final Rule, *Federal Register, 56(126)*, 30266-30281, July 1, 1991.

U.S. Environmental Protection Agency (1991d), 40 CFR Parts 141, and 142; National Primary Drinking Water Regulations; Radionuclides; Proposed Rule, *Federal Register, 56(138)*, 33050-33127, July 18, 1991.

U.S. Environmental Protection Agency (1992), 40 CFR Parts 141, and 142; National Primary Drinking Water Regulations; Synthetic Organic Chemicals and Inorganic Chemicals; Final Rule, *Federal Register, 57(138)*, 31776-31849, July 17, 1992.

U.S. Environmental Protection Agency (1994a), 40 CFR Parts 141, and 142; National Primary Drinking Water Regulations; Disinfectants and Disinfection Byproducts; Proposed Rule, *Federal Register, 59(145)*, 38668-38829, July 29, 1994.

U.S. Environmental Protection Agency (1994b), 40 CFR Parts 141, and 142; National Primary Drinking Water Regulations; Enhanced Surface Water Treatment Requirements; Proposed Rule, *Federal Register, 59(145)*, 38832-38858, July 29, 1994.

U.S. Environmental Protection Agency (1994c), 40 CFR Parts 141, 142, and 143; National Primary Drinking Water Regulations—Sulfate; Proposed Rule, *Federal Register, 59(243)*, 65578-65604, December 20, 1994.

U.S. EPA, see U.S. Environmental Protection Agency.

3

Amenability of Water Quality and System Operating Parameters

An initial list of selected national primary and secondary drinking water quality parameters (or parameter categories) and operating parameters is presented in Table 3-1. Amenability ratings are given for remote monitoring and control within the framework of specific treatment technologies used by small treatment systems. Additional parameters may be selected from the complete list after the specific treatment process and water quality have been identified for field-testing of the prototype RTS. The MCLs given in Tables 2-4 through 2-6 can be used as starting points to determine the practical ranges for product water quality monitoring by remote-sensing devices. In contrast, raw water quality may vary significantly from location to location or even seasonally within the same location. Accordingly, the monitoring of raw water quality by remote-sensing devices should allow for greater variability. A summary of applicable ranges and analytical techniques available for the determination of target water quality parameters also is presented in Table 3-1. Devices with various degrees of automation are commercially available for many of the methods presented in Table 3-1. The COTS devices are discussed in Chapters 4 and 5.

Table 3-2 summarizes recommended remote-sensing monitoring and control needs for selected water quality and operating parameters within the framework of the various treatment technologies used by small treatment systems. Depending on the treatment technology under consideration, monitoring and/or control of each parameter might be applicable, either generally or for specific types of water as discussed subsequently in this chapter.

TABLE 3-1. Summary of Methods, Ranges for the Determination of Target Water Quality and Operating Parameters, and Amenability to Process Control and Monitoring

Parameter Type	Parameter	Methods	Range	Amenability
Water Quality	Microbial Contaminants	Culture Molecular probes	0-100 cyst or oocyst/L 0-10^7 virus or bacteria cell/L	L[(a)]
	Turbidity	Nephelometry	1-100 NTUs	M[(b)]-H[(c)]
	pH	Electrometry	4 -10 pH units	H
	Chemical Dose/Residual (Free Chlorine)	Iodometry (total) Amperometry (free, combined) N,N-diethyl-*p*-phenylenediamine (DPD) colorimetry Syringaldazine colorimetry (free)	0-4 mg/L expected. Monitor 0-10 mg/L	H
	Temperature	Thermocouples Thermistors Infrared	Surface water: 0-40°C Groundwater: 15-20°C	H
	Total Organic Carbon (TOC)	Oxidation (combustion, wet, or persulfate ultraviolet) Detection (infrared or flame ionization)	0.5-10 mg/L	L
	Trace Metals (Iron and Manganese)	Atomic absorption spectrophotometry Inductively coupled plasma Phenanthroline colorimetry (iron) Persulfate colorimetry (manganese)	Groundwater 0-20 mg/L	L
	Anions/Cations (Fluoride/Nitrate)	Electrometry Colorimetry	Fluoride 0-10 mg/L Nitrate 0-50 mg/L as N	M-H
	Color	Platinum-cobalt colorimetry Spectrophotometry Tristimulus filter	0-100 color units	M-L
	Hardness (Calcium Hardness)	Ethylenediaminetetraacetic acid (EDTA) titration Atomic absorption spectrophotometry Inductively coupled plasma	0-150 mg/L as Ca^{2+}	M-L
	Total Dissolved Solids (TDS)	Gravimetry Conductivity	0-500 mg/L to 0-5,000 mg/L	M-H

TABLE 3-1. (continued)

Parameter Type	Parameter	Methods	Range	Amenability
Water Quality (cont'd)	Particle Counts	Electrical sensing zone Light blockage Light scattering	0-10^{10}/L	M
System Operation	Flowrate	Mechanical turbine/impeller flowmeters Rotameters Vortex shedding Magnetic differential pressure Ultrasonic Doppler	Varies based on size of plant	H
	Pressure	Pressure transducers Differential pressure transducers	0-400 psig[d]	H
	Electricity Usage	Electronic home meters	Varies with application	H
	Volume	Mechanical or electronic totalizers	Varies with application	H
	System Status (e.g., On/Off, Backwash)	Contact relays	NA	H
	Time	Digital counter	Varies with application	H
	Chemical Feed Levels	Ultrasonic Acoustic Conductance electrodes Float switches Capacitor switches Strain gauge	Varies with application	H

(a) L = Low amenability.
(b) M = Moderate amenability.
(c) H = High amenability.
(d) psig = pounds per square inch, gage.

TABLE 3-2. Remote-Sensing Monitoring/Control Needs for Selected Water Quality/Operating Parameters within Treatment Technologies Used by Small Drinking Water Systems

		Water Quality Parameters												Operating Parameters							
Source Water	Technology	Microbial	Turbidity	pH	Chemical Dose/Residual	Temperature	TOC	Fe and Mn	Fluoride	Color	Hardness	TDS	Particles	Flowrate	Pressure	Electric Usage	Finished Water	Wastewater Produced	Status (on-off, backwash, etc.)	Hours	Feed Chemical Levels
GW	Air Stripping	NA	NA	M,C	M,C	M	NA	M,C	NA	NA	M	NA	NA	M	NA	M	M	NA	M	NA	M
	Oxidation/Filtration	TC	M,C	M,C	M,C	M	M	M,C	NA	M,C	NA	NA	M	M,C	M,C	M	M	M	M,C	M	M,C
	Ion Exchange	TC	NA	M,C	M	M	NA	M	M	NA	M,C	NA	NA	M,C	NA	M	M	M,C	M	M,C	NA
	Activated Alumina	TC	NA	M,C	M,C	M	NA	NA	M,C	NA	NA	NA	NA	M	NA	M	M	M,C	M	M,C	M,C
SW	Coagulation/Filtration	TC	M	M,C	M	M	M	NA	NA	M	NA	NA	M	M,C	M,C	M	M	M	M,C	NA	M,C
	Dissolved Air Flotation	NA	M	M,C	M	M	M	NA	NA	M	NA	NA	M	M	NA	M	M	M	M,C	NA	NA
	Diatomaceous Earth Filtr.	TC	M	M	NA	M	M	NA	NA	M	NA	NA	M	M,C	M,C	M	M	M	M,C	NA	M,C
	Slow Sand Filtration	TC	M	M	NA	M	M	NA	NA	M	NA	NA	M	M,C	M,C	M	M	M	M,C	NA	M,C
	Bag/Cartridge Filtration	TC	M	M	NA	M	M	NA	NA	M	NA	NA	M	M,C	M,C	M	M	M	M,C	NA	M,C
GW and SW	Disinfection	CT	M	M,C	M,C	M	M	M	NA	NA	NA	NA	NA	M,C	NA	M	M	NA	M	NA	M,C
	Corrosion Control	NA	M	M,C	C	M	M	M	NA	NA	M,C	M	NA	M	NA	M	M	NA	M	NA	M,C
	Membrane Filtration	TC	M	M,C	M	M	M	NA	NA	M	NA	NA	M	M,C	M,C	M	M	M,C	M,C	NA	M,C
	RO/NF	TC	NA	M,C	M,C	M	M	M	M	M	M,C	M,C	NA	M,C	M,C	M	M	M,C	M,C	NA	M,C
	Electrodialysis	NA	M	M,C	NA	M	NA	M	M	NA	M	M,C	NA	M,C	NA	M	M	M,C	M,C	NA	NA
	Adsorption	TC	M	M	NA	M	M	M	NA	M	NA	NA	NA	M	NA	M	M	NA	M,C	NA	NA
	Lime Softening	NA	M	M,C	M,C	M	NA	M	M	M	M,C	NA	NA	M	NA	M	M	NA	M	NA	M,C

GW = Groundwater.
SW = Surface water.
NA = Generally not applicable with respect to parameter monitoring or control by remote sensing within specific treatment technology.
TC = Treatment credit in terms of minimum log removals (e.g., 2.0 log removal of *Giardia lamblia* cysts in slow-sand filtration units).
CT = Microbial contaminant control compliance currently demonstrated with *CT* approach (i.e., product of average disinfectant residual concentration and t_{10} contact time).
M = Parameter monitoring by remote-sensing system required/desirable generally or for certain water qualities.
C = Parameter control integral to overall remote-sensing process control system, generally or for certain water qualities.

3.1 Water Quality Parameters

Twelve water quality parameters (or parameter categories) are included in Tables 3-1 and 3-2. The parameters are discussed in Sections 3.1.1 through 3.1.12 for their amenability to remote monitoring and control within the framework of specific treatment technologies used by small treatment systems.

3.1.1 Microbial Contaminants

Pathogenic microorganisms of interest to drinking water include viruses, bacteria, and protozoan cysts/oocysts. The regulatory approach under the SWTR and expected for the ESWTR and GWDR is based on requiring water utilities to demonstrate adequate disinfection capacity. Existing disinfection data support the approach that disinfection systems designed to inactivate the more resistant microorganisms, i.e., protozoa and viruses, should provide excess inactivation capacity for bacteria. The *CT* approach presented in Chapter 2 is selected to demonstrate regulatory compliance because of the current lack of analytical techniques for the detection and viability assessment of microorganisms in natural and treated water. However, various techniques currently are under development for the in situ detection of specific pathogens in addition to biological and nonbiological surrogate indicators. A review of these techniques is presented in Chapter 8.

3.1.1.1 *CT* Method

Small systems treating surface water can use the *CT* approach to demonstrate compliance with the SWTR. It is anticipated that the *CT* approach can be used by water utilities, including small systems treating groundwater, to also demonstrate compliance with disinfection requirements under the forthcoming GWDR. A remote-sensing system should be designed in a manner similar to that described in Chapter 2 for the groundwater disinfection plant. Regardless of the type of treatment process under consideration, the system should include disinfectant addition and residual monitoring capabilities. The automation scheme should be based on maintaining a *CT* above the minimum required value at the specific pH and temperature (see Tables 2-7 and 2-8). Depending on the actual inactivation efficiency required, which in turn depends on the type of treatment plant under consideration, the limiting microorganism could be *G. lamblia* cysts (68 to 99% inactivation) or viruses (99 to 99.99% inactivation). It is expected that, under the ESWTR, *Cryptosporidium parvum* (*C. parvum*) oocysts will become the limiting microorganisms for which surface water disinfection plants should be designed. The *CT* requirements for *C. parvum* oocysts are not yet available.

3.1.1.2 Tracer Tests

Automation and remote-monitoring systems for the control of disinfection efficiency should be designed to maximize the disinfection credit of specific water treatment systems. In general, the most advantageous approach would be to utilize not only the portion of the treatment plant after disinfectant addition but also the portion of the distribution system prior to reaching the first customer. This approach may require the performance of tracer tests in which the tracer chemical is added at the disinfectant addition point and the tracer curve is determined at the distribution system location at which the residual is measured.

3.1.2 Turbidity

Typical values for turbidity depend on the source water and range from 0 to 1,000 NTUs. Unless influenced by surface water or wastewater, groundwater typically has consistently low turbidity levels. However, this condition might change in the effluent from certain types of malfunctioning groundwater treatment plants designed to remove hardness, iron, and manganese. The turbidity of surface water can vary widely even within the same location. Accordingly, remote monitoring of raw water turbidity is important primarily for surface water and, in turn, for groundwater under the influence of surface water.

The SWTR specifies a monthly average maximum turbidity of 1 NTU and an absolute maximum of 5 NTU for finished water produced by systems treating either surface water or groundwater under the influence of surface water. Turbidity requirements for finished groundwater, not yet defined, will be addressed under the GWDR.

Turbidity may be removed by filtration and/or coagulation/flocculation/sedimentation processes. The remote monitoring and control needs for these processes are discussed in Sections 3.1.2.1 and 3.1.2.2.

3.1.2.1 Filters and Membranes

Turbidity measurements should be taken before and after the filter and membrane units. Turbidity changes in the product water may indicate breakthrough of conventional filters or a loss of integrity by a membrane system. Measurements before and after these units help a remote operator determine if the plant is running correctly. Bag and cartridge filters are appropriate for high-quality source water. Their removal efficiency for turbidity is approximately 50%. Accordingly, bag and cartridge filters should not be used to treat water with turbidity greater than 2 NTUs.

3.1.2.2 Coagulation/Flocculation/Sedimentation Processes

Surface water typically is characterized by turbidity fluctuations. These fluctuations create dynamic conditions in a treatment plant that affect the operation of coagulation/flocculation processes. Therefore, coagulation/flocculation/sedimentation processes are not recommended for remote operation.

3.1.3 pH

Typical values for pH depend on the source water and could range from 5 to 11.5 pH units. However, the pH of a specific surface water or groundwater tends to vary within a narrower range. Accordingly, online pH control including a probe calibration system should be selected with the pH of the source water in mind. The pH of treatment plants using chlorine for disinfection must be measured at least once per day at each chlorine residual sampling point.

Monitoring and control of pH are especially important for processes such as breakpoint chlorination, corrosion control, and membrane softening. The specific needs for these processes are described in Sections 3.1.3.1 through 3.1.3.3.

3.1.3.1 Breakpoint Chlorination

Groundwater, primarily in rural areas, tends to be contaminated seasonally with ammonia nitrogen from crop fertilizers. To achieve a stable residual of stronger disinfectant free chlorine, the formation of chloramines must be avoided by practicing breakpoint chlorination. Breakpoint chlorination is the process in which free chlorine is added at levels that result in the oxidation of ammonia nitrogen primarily to nitrogen gas. The chlorine dose should satisfy the ammonia oxidation demand, i.e., chlorine:ammonia ratios between 1.5 and 2.0 on a molar basis, or between 7.5 and 10 on a weight basis (mg chlorine as Cl_2/mg ammonia as N). Additional chlorine doses in excess of that consumed by ammonia oxidation would be required to achieve a given residual corresponding to the system *CT* requirement. The rate of the breakpoint chlorination reaction is fastest at pH levels in the range of approximately 7 to 8, and it tends to slow down at values below and above this optimum range. Monitoring and possibly control of the pH thus are critical for optimization of breakpoint chlorination.

3.1.3.2 Corrosion Control

Low pH produces greater corrosion potential in a water distribution system. Some corrosion inhibitors, such as limestone, will increase the pH and alkalinity. Others, such as orthophosphates and polyphosphates, may require a pH adjustment step. Orthophosphates work best at a pH ranging from 7.2 to 7.8.

3.1.3.3 Membrane Softening

NF and low-pressure RO membranes can be used for softening, i.e., removing divalent cations, primarily calcium and magnesium. As discussed in Chapter 2, the reject water from the membrane system is concentrated, and consequently, the concentrations of ions present in solution might result in supersaturation and detrimental precipitation of sparingly soluble compounds such as calcium carbonate. Calcium carbonate precipitation produces membrane fouling and ultimately process failure. This problem can be avoided by decreasing the pH of the water upstream from the membrane system to ensure that the acidified solution remains undersaturated with respect to calcium carbonate. The acid dose requirement is a function of raw water pH and alkalinity. The pH of the product water from the membrane system must be increased, typically to pH 8 to 9, to prevent corrosion of the distribution system components. The dose of caustic to be added would depend on the pH and the carbon dioxide/carbonic acid and/or alkalinity concentration of the product water.

3.1.4 Chemical Dose/Residual Concentration

With the exception of groundwater treatment plants in certain states, small systems must add disinfectants and measure the corresponding residual after contact time. The disinfectant residual is limited by the maximum residual disinfectant level (MRDL) as an upper limit, 4 mg/L as Cl_2 for both free chlorine and chloramine (see Table 2-6). The residual also has a lower limit set by the requirement of maintaining a minimum residual concentration in the distribution system of 0.2 mg/L as Cl_2, or providing the minimum *CT* required as discussed previously. Ozone is very reactive and does not provide a stable residual, and chlorine dioxide disinfection is a more complex process not recommended for small systems (NRC, 1997).

3.1.5 Temperature

The temperatures of surface water in the United States can range from 0 to 40°C, both regionally and seasonally. In contrast, a specific groundwater tends to have a constant temperature in the range of 15 to 20°C at the well. Temperature is an important parameter to monitor during the treatment of drinking water. Temperature has an effect on most biological, chemical, and physical processes. In general, the overall or apparent rate of most chemical and biochemical reactions tends to decrease with temperature. Temperature also has a strong effect on the volatility of contaminants, for example, the efficiency of air stripping decreases rapidly with decreasing temperature. Additional examples justifying the need for monitoring temperature are presented below.

3.1.5.1 Turbidity Monitoring

As temperature decreases, chemicals added to aid coagulation become less effective. Temperature measurements would allow a remote operator to adjust the feed rate of these chemicals to adjust for their decreased effectiveness. In addition, if it is found that a newly installed flocculation chamber is not performing as expected, temperature data may assist in finding the problem. Viscosity, a design parameter of flocculation chambers, is a function of temperature. A package plant designed for conditions much warmer than those at the place of installation may be underdesigned.

3.1.5.2 Disinfection

Temperature has a strong effect on the disinfection efficiency attained with all common drinking water disinfectants. This issue was discussed previously in this book in terms of the effect of temperature on *CT* parameters (see Tables 2-7 and 2-8).

3.1.5.3 Membranes

Temperature affects the operation of membrane treatment units. The permeation of water across the membrane decreases with decreasing temperature. This decrease in water productivity must be balanced by increasing the operating pressure of the membrane system with the goal of maintaining a constant product water flow.

3.1.5.4 Adsorption

The effects of temperature on physical sorption and diffusion are not very significant. However, these minor effects are included here for completeness.

3.1.6 Total Organic Carbon

The total organic carbon (TOC) content of source waters can range from 0 to 20 mg/L, and the national average is approximately 3 mg/L. In general, the levels of TOC found in surface water tend to be greater than those encountered in groundwater.

TOC in the source water is of concern for two reasons. First, waters containing TOC at the disinfection step are likely to form DBPs of public health concern. DBP regulations have been proposed under the Disinfectants/Disinfection Byproduct Rule (see Table 2-4). Second, TOC provides carbon sources for microorganisms. The biodegradation of these carbon sources takes up oxygen and may allow the water to become anaerobic. TOC, however, may not need to be monitored during treatment. Two processes that would benefit from TOC monitoring are discussed below.

3.1.6.1 Coagulation

Water containing higher levels of TOC will have a higher coagulant demand. Treatment systems using surface waters with fluctuating TOC levels are difficult to control and are not recommended for remote monitoring.

3.1.6.2 Chlorination

Water with high levels of TOC tends to exert high disinfectant demand, thus requiring greater disinfectant dosages. Free chlorine reacts with TOC compounds producing DBP including trihalomethanes (THMs), haloacetic acids (HAAs), and chloral hydrate for which regulations have been proposed (see Table 2-4). However, treatment systems using surface waters as intakes are difficult to control, and monitoring of DBP formation requires sophisticated laboratory instrumentation. Therefore, remote monitoring of TOC is not recommended for small treatment systems.

3.1.7 Trace Metals

Trace metals of public health concern that are encountered in surface water and groundwater and their corresponding MCLs are summarized in Table 2-6. Remote control of treatment plants designed to remove these compounds should include online probes, if available, calibrated to demonstrate compliance by detecting levels in the order of magnitude of the corresponding MCL in the finished water. The concentration in the raw water could range widely, with maximum values typically up to several milligrams per liter (mg/L).

Iron and manganese are the most common trace metals encountered in groundwater at concentrations up to 20 mg/L each. Although there is no national primary drinking water regulation for iron and manganese, the aesthetic and other secondary problems created require that treatment for these compounds be provided if they are present in the source water. Iron and manganese are relatively soluble when present in their reduced oxidation state, the predominant forms in anoxic groundwater. However, in the presence of oxygen they tend to oxidize into forms with very low solubility, thus creating the unwanted classic red water and staining problems. Iron can be removed from groundwater by chemical oxidation with dissolved oxygen, free chlorine, or permanganate, the latter two chemicals generally providing greater oxidation efficiency. The rate of oxidation with dissolved oxygen is faster at relatively higher pH and generally requires a pH greater than 7.5 or 8.0. The solids resulting from the oxidation/precipitation step can be removed in a filter. The oxidation of manganese is more complex and presents a greater challenge for automation. Separation processes used to remove trace metals include ion exchange and membrane processes. Co-precipitation also can be achieved by using lime soda in softening units.

3.1.8 Anions/Cations

Most of the inorganic contaminants, including the trace metals discussed in Section 3.1.7, regulated under the SDWA (see Table 2-6), occur in natural water in the form of anions or cations. Among these compounds, fluoride and nitrate are of particular interest. The concentration of fluoride in natural water could be as high as 20 mg/L, or greater than the current MCL of 4 mg/L. In surface water and groundwater, nitrate levels as N resulting from the application of agricultural fertilizers could be several times in excess of the MCL of 10 mg/L. Treatment processes to remove these contaminants include ion exchange, RO, and NF. Activated alumina has been applied successfully for the removal of fluoride. The remote monitoring and control needs for these processes are discussed in Sections 3.1.8.1 through 3.1.8.3.

3.1.8.1 Activated Alumina

Activated alumina is useful for removing negatively charged ions at pH below 9.5. When treated with an acid solution, pH 5.5, activated alumina is highly selective for fluoride, selenium, and arsenic. The spent media may be disposed of or regenerated with caustic solutions. Commercial activated alumina units can be automated to treat a given volume of water prior to disposal or regeneration.

3.1.8.2 Ion Exchange

Ion exchange can remove anionic or cationic species depending on the resin selected. Anion exchange resins can remove species such as nitrate, carbonate, dichromate, fluoride, selenate, and selenite and replace these ions with hydroxide or chloride ions. Cation exchange resins can remove magnesium, calcium, strontium, ammonium ion, barium, radium, cadmium, lead, and trivalent chromium and replace the ions with protons, sodium, or potassium.

Ion exchange resins need to be regenerated with brines of chemicals containing the original exchange ion before exhausting the resin exchange capacity. Commercial ion-exchange units typically are automated to treat a given volume of water prior to regeneration. However, meeting the MCL for specific contaminants might require process automation based on starting regeneration when reaching a target concentration at the ion-exchange column effluent, which in turn requires the use of an online probe.

3.1.8.3 Membranes

NF and RO membranes can remove most of the regulated inorganic contaminants listed in Table 2-6 at levels in the range of 90 to 99.99%, depending on the type of membrane selected and the target contaminant of interest. The treatment of groundwater could be accomplished with a unit similar to that presented in Figure 2-2 after including the corresponding probe for monitoring the specific contaminant of interest. The treatment of surface water with NF or RO membranes requires more pretreatment than groundwater applications for suspended solids removal. The maximum turbidity allowed in the feed to NF and RO membranes typically is 1 NTU. Examples of pretreatment processes used with NF and RO membranes are UF and MF membrane technologies and various media filtration processes.

3.1.9 Color

Color in source water has a nationwide range of 0 to 400 color units. This range is too wide to effectively examine the problem of color in drinking water by a standard approach. Color, instead, should be addressed on a case-by-case basis. Most notably, groundwater withdrawn from certain shallow aquifers or swamps tends to have high color levels in the upper end of the range provided.

3.1.9.1 Coagulation

Color in water is related to the TOC content, and is removed by the same coagulation mechanism as TOC. Color is not a control parameter for this process. Coagulation processes are not recommended for remote monitoring.

3.1.9.2 Nanofiltration and Reverse Osmosis

NF and RO membranes can be used to remove TOC associated with color. Typical color removal efficiencies achieved in NF and RO plants range from 90 to 99%. Color is not, however, a control parameter for this process.

3.1.9.3 Chemical Oxidation

Chemical oxidation can be used to break the double bonds responsible for the color in the TOC molecules. The resulting water is no longer colored, but it has the same level of TOC. The level of color in the water will determine the chemical dose. Color should be measured before and after this process for process control.

3.1.10 Hardness

Hardness is the measure of calcium and magnesium in the water. The greater the value of hardness, the greater the potential for the water to cause problems with precipitation in the distribution system, home water heaters, and industrial process piping. According to hardness content, water can be classified as soft (total hardness of less than 75 mg/L as $CaCO_3$), moderately hard (75 to 150 mg/L as $CaCO_3$), hard (150 to 300 mg/L as $CaCO_3$), and very hard (greater than 300 mg/L as $CaCO_3$). The typical range of hardness in raw waters depends on the geology of the region. Regions with more calcium and magnesium in the soil will have source waters with higher hardness. Note that hardness is not a process control parameter but a parameter to be monitored in the finished water.

3.1.10.1 Lime Softening

Lime softening is a process for removing hardness by precipitation. The amount of lime added depends on the hardness of the water. The chemistry of lime softening is complex. This process is not recommended for small systems or remote control.

3.1.10.2 Ion Exchange

Ion exchange removes hardness by substituting the calcium and magnesium in the water for another ion, usually sodium, as discussed in Section 3.1.8.2.

3.1.10.3 Nanofiltration and Reverse Osmosis

NF and RO membranes efficiently remove hardness components. The layout of a treatment plant using NF and RO is shown in Figure 2-2.

3.1.11 Total Dissolved Solids

The concentration of total dissolved solids (TDS) in certain types of groundwater and surface water can exceed 500 mg/L. For example, some of the aquifers heavily used in populated coastal areas of Florida and California are being contaminated by intrusion of seawater. These waters require some level of desalination.

The concentration of TDS can be reduced with RO membrane processes. The removal rate depends on the membrane and the valence of the ions in question. The TDS concentration needs to be monitored, but it should not be considered a process control parameter.

3.1.12 Particle Counts

The characterization of particles present in raw and treated water streams can serve as an indicator of the performance of water treatment processes, including the various chemical conditioning and media filtration processes discussed in this book. Automated particle counters are commercially available for remote monitoring. An attractive advantage of particle counters over turbidimeters is that particle counters could be used to characterize particle size distributions, which allow the monitoring of particles in the size range of target pathogens. However, particle counters currently are not useful for process control because of the complex relationship among treatment efficiency, process operating conditions, and the physico-chemical principles involved.

3.2 System Operating Parameters

In addition to monitoring both raw and product water quality, it is important to monitor and control operational parameters to ensure safe and optimal operations. The parameters to be monitored and/or controlled include flowrate, pressure, utility usage, volumetric readings of both production and reject water, system status, hours of operation, and chemical feed levels for both plant operations and sensor calibration requirements, as shown in Tables 3-1 and 3-2. Sections 3.2.1 through 3.2.8 present information on these parameters.

3.2.1 Flowrate

Flowrate is a process parameter that indicates production levels and pumping efficiencies. The groundwater disinfection plant described in Chapter 2 would require the use of at least two flowmeter/totalizer units: one ahead of the system's feed pump and the other on the backside of the system where the product water is delivered to the distribution system. By remotely monitoring the flowrate at these two points, it would be possible to monitor the integrity of the piping system and pump performance. Furthermore, the control system should include the monitoring of flowrate and t_{10} contact time. If the source water is of consistent quality, the disinfectant dose can be determined by the flowrate.

If membrane processes are used, membrane plants typically are designed for fixed feed and product water flowrates. A controller would use flowrate data to adjust the pressure across the membrane system as the membrane surfaces undergo unavoidable fouling.

Flowrate/totalizer systems also may be used to monitor the volume of wastewater production. Monitoring wastewater volume is critical if discharge has to be in compliance with permit requirements.

3.2.2 Pressure

Two types of pressure measurements may be incorporated into the RTS, i.e., a single-point measurement of either line or vessel pressure and the differential pressure across a plant function. The single pressure measurement could be used as an indicator of pump performance and transfer pipe integrity. If the line pressure downstream of a pump is not within expected operating ranges, either pump or pipe failure could have occurred. Pressure monitoring also may be applied to storage systems as an alternative to level measurements. For simple water treatment operations where filtration is not employed, it is expected that pressure sensors covering a range from 0 to 100 psi would be appropriate.

Pressure monitoring is essential for process controls involving filtration or membrane operations, as discussed in Chapter 2. In the case of single-speed centrifugal pumps, pressures in the range of 100 to 400 psi can be expected just after the NF/RO pump. The pressure just upstream of the membrane system (downstream from the pressure-control valve shown in Figure 2-14) can be expected to be between 50 to 200 psi for clean membranes and when the water temperature is relatively high. This pressure range might have to be increased to between 100 to 400 psi for fouled membranes and when the temperature of the water is relatively low. Periodic membrane cleaning should be done after exceeding the maximum pressure within the range.

Differential pressure is an important parameter for conventional filtration processes when treating either groundwater or surface water. However, these processes are not recommended for remote control.

3.2.3 Electricity Usage

Electricity usage, although not providing a parameter for process control or water quality, does monitor a utility data need. Readily available electrical totalizers could be integrated into an RTS.

3.2.4 Finished Water Production

Measurement of finished water produced has an important implication for water quality in the distribution system. The data should be compared to actual distribution consumption through billing records to determine if losses were occurring in the distribution system. Excessive losses may correspond not only to inefficiency but also to potential exposure of the water to contamination.

3.2.5 Wastewater Production

Some wastewater is expected to be produced from small treatment systems that include treatment efforts beyond simple disinfection. Wastewater from membrane treatment is at a relatively fixed flowrate. Measurements of this flowrate and the total volume produced, although not useful for process control or remote monitoring of water quality, may provide insight into the performance of the membrane.

Online calibration for quality assurance of any RTS sensors is expected to result in only a small volume of wastestream. This wastestream most likely would be retained in a vessel for proper disposal.

3.2.6 System Status

System status data, or on/off flags for equipment, are useful data for remote process control. These data would indicate if expected operational conditions exist or if a mechanical problem has occurred. For example, if the chlorine residual concentration is within an acceptable range, the chlorine delivery mechanics are working properly. This information becomes useful when the chlorine residual drops below the target level. If this happens, the operator should be able to rule out the status of the chlorine injection process and move on to issues such as changing water quality and increasing demand.

3.2.7 Hours of Operation

Hours of operation for mechanical devices such as pumps should be measured so that maintenance can be scheduled. Measurements can be accomplished with electrical totalizers or a temporal on/off flag tied to the RTS internal clock.

3.2.8 Level Measurements

Level measurements of chemical feed and production water and their monitoring are expected to be included in the RTS. These parameters are discussed in Sections 3.2.8.1 and 3.2.8.2.

3.2.8.1 Chemical Feed Levels

Tanks or vessels containing chemical feedstocks or calibration mixtures are expected to be fitted with level indicators. Level sensors will be integrated with a signal output device that can be accessed remotely for replenishment needs. Level sensors also may be interfaced with an alarm system that would contact the host computer to indicate a need for refill. After this alarm process has been validated, calls directly to either a pager or an operator's home phone may be included. This will be done only after thorough confirmation that the alarm system is reliable to eliminate possible false alarms.

3.2.8.2 Production Water Tank Levels

Water storage tanks may be fitted with a sensor to allow remote monitoring of the volume available for distribution. Distribution system storage tanks can hold from as much as a 3-day supply to as little as a 1- to 2-hour supply, depending on the water treatment system. When the level in the storage tank drops to a specific volume, the treatment operation goes on line to replenish water that has been used. An RTS level sensor or pressure transducer would monitor this water consumption. If the treatment system fails to respond to a low-supply situation, this would cause an alarm at either the RTS host computer or a specified operator response location.

3.3 Reference

National Research Council (1997), *Safe Water from Every Tap. Improving Water Service to Small Communities*, Committee on Small Water Supply Systems, Water Science and Technology Board, Commission on Geosciences, Environment, and Resources, National Research Council, National Academic Press, Washington, D.C.

NRC, see National Research Council.

4

COTS Water Quality Monitoring and Control Devices

This chapter provides the results of the literature search for the COTS availability of each of the 12 water quality parameters discussed in Chapter 3. Spreadsheets are used to present the operating principles, performance specifications, calibration/maintenance requirements, dimensions, and costs of the COTS devices identified for each of the parameters. Individual data sheets for all COTS devices discussed in this chapter are presented in Appendix B.

4.1 Microbial Contaminant Monitors

As discussed in Section 3.1.1, pathogenic microorganisms in drinking water currently are or will be regulated under the SWTR, ESWTR, and/or GWDR. No COTS monitoring/sensing devices have been identified as capable of detecting and assessing the viability of microorganisms in natural and treated water. Particle-counting instruments (Section 4.12) that can selectively monitor the particle-size range associated with microorganisms may be the only options for this water quality parameter at this time. Several techniques currently are being developed for the online detection of specific pathogens and biological and non-biological surrogate indicators. A review of these methods is included in Chapter 9.

4.2 Turbidity Sensors

A review of manufacturer's literature resulted in the identification of eight COTS turbidity-sensing devices (Table 4-1). A basic nephelometric turbidity sensor consists of a light source and a photodiode that detects the scattered light in the presence of turbidity at 90° to the incident light (see Figure 4-1). The light detected is converted to a signal and linearized by an electronic analyzer. The amplitude of the signal indicates the level of turbidity in the sample and is reported in NTUs.

The second type of light dispersion turbidity sensor utilizes an infrared (IR) light-emitting diode (LED) and two photoelectric detectors. The first detector measures the scattered light,

TABLE 4-1. Features of COTS Turbidity Sensors

Manufacturer/ Model	Rosemount A2120-FT-M	Hach 1720D	Honeywell	GLI International 95T/8220
Technology	Light scattered at 90° detected by photodiode	Light scattered at 90° detected by submerged photocell	Infrared LED and two detectors (transmissive and scattered)	Ratiometric four-beam method
Range/ Sensitivity	0.000-200 NTUs; resolution is 0.001 NTU below 1 NTU	0-100 NTUs; resolution is 0.001 NTU	0-4000 NTUs for turbidity	0.000-100.0 NTUs with autoranging; sensitivity is 0.001 NTU
Accuracy	±2% of reading from 0-60 NTUs; ±5% of reading from 60-200 NTUs	±2% of reading, or ±0.020 NTU, 0-30 NTUs; ±5% of reading, 30-100 NTUs	See manufacturer's curve	±2% of reading, all ranges
External Dimensions	Flowthrough chamber 4.38" W x 8.25" H x 4.38" D; 0.5" NPTM inlet and outlet connections	9 x 9 x 7" each for AquaTrend interface and power supply; 10 x 12 x 16" for body and cap	Mounting dimension is 1.25" diameter x 2.39"; cap is 0.71" OD x 2.75"	12.5" W x 9.7" H x 6.6" D
Calibration/ Maintenance Requirements	Calibrated with Formazin or AEPA-1 polymer standards; also secondary standard for quick calibration checks	Requires minimal maintenance; calibration based on Formazin standards	NR	Periodic verification of calibration using turbid glass cube; occasional cleaning of flow chamber; LED light lasts 10 to 12 yr
Cost	$1,776 for A2100 sensor; $823 for T1054 analyzer	$950 for turbidimeter and power supply; $1,795 also includes AquaTrend interface	$90 without flowthrough cap; $100 with flowthrough cap	$950 for analyzer; $1,050 for sensor
Comments	Alarm relays; flowthrough or debubbling chamber; integral preamplifier	Two setpoint alarms	Also measures conductivity and temperature; with software and microprocessor	Designed for lower turbidity

TABLE 4-1. (continued)

Manufacturer/ Model	GLI International 95T/8224T	Global Water WQ700	OMEGA TRCN95	Environmental Sensors 2600
Technology	Ratiometric four-beam method	Incandescent light and 90° light scattering	Four-beam ratiometric measurement	Backscatter nephelometer
Range/ Sensitivity	0.00-1,000 NTUs with autoranging; sensitivity is 0.01 NTU	0-200 NTUs; optional 0-2,000 NTUs or 0-10 NTUs	0.001-100 NTUs; sensitivity is 0.001 NTU	500, 1000, 2000 FTUs
Accuracy	±5% of range	± 0.1 NTU for 0-200 NTUs	±2% of reading	2% of full scale
External Dimensions	12.5" W x 9.7" H x 6.6" D	Body is 1.1" x 6"; shield is 1.7 x 4"; overall length is 8"	Sensor is 17.5" H x 13.8" W x 7.6" D; analyzer is 12.5" H x 9.7" W x 6.6" D	7.875" L x 1.25" diameter
Calibration/ Maintenance Requirements	Periodic calibration using turbid glass cube; LED light lasts 10 to 12 yr	Incandescent lamp has 20-yr continuous usage	Occasional cleaning of sample chamber baffle and glass window; calibration with Formazin or optical calibration cube	NR
Cost	$950 for analyzer; $2,200 for sensor	$800 for sensor; $285 for LCD; $185 for flowthrough chamber	$1,995; $395 for calibration cube	$1,200
Comments	Designed for higher turbidity	Rugged housing	Automatic color compensation; bubble-rejecting chamber; warning and failure signals	NR

FTU = Formazin nephelometric turbidity unit.
LCD = liquid crystal display.
LED = light-emitting diode.
NPTM = normal pipe thread, male.
NR = not reported.
OD = outside diameter.

whereas the second detects transmissive light. As scatter increases for a water sample containing high levels of turbidity, transmission correspondingly decreases. This response is inverted for water with less turbidity. The turbidity, therefore, is measured based on the ratio of scattered light to transmitted light. The optical design of a Hach ratio turbidimeter is shown in Figure 4-2. The signal generated by the sensor is reported in NTUs.

The third type of light dispersion turbidity sensor uses a four-beam method consisting of two light sources and two detectors. The sources and detectors are positioned to obtain multiple signals. The light sources are activated in series along with the detectors, providing both active and reference measurements for each light source. The turbidity value is calculated based on a ratiometric algorithm of the four readings. The configuration of the modulated four-beam method is shown in Figure 4-3.

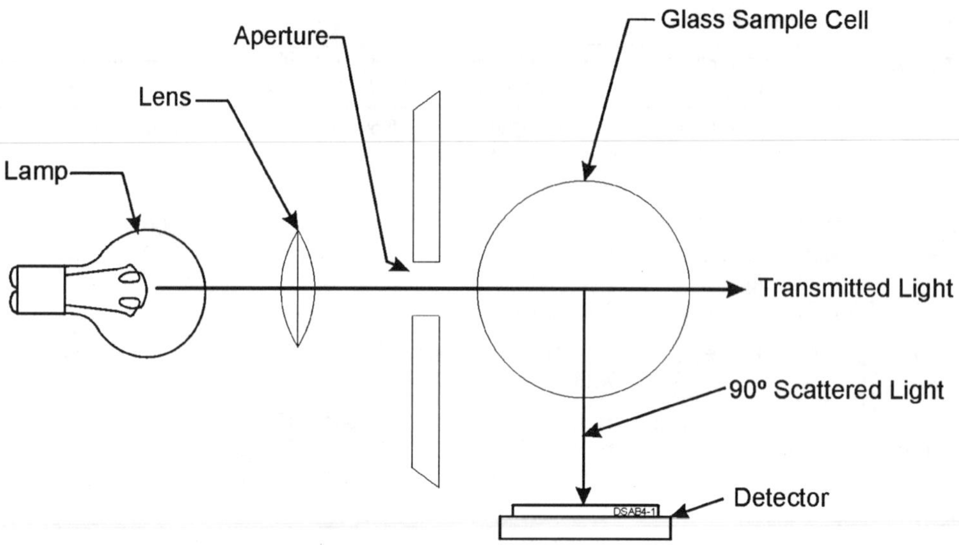

FIGURE 4-1. A basic nephelometric turbidity sensor that detects scattered light at 90° to the incident beam. (Adopted from M.J. Sadar, *Understanding Turbidity Science*, Technical Information Series Booklet No. 11. Hach Company, ©1996. Used with permission.)

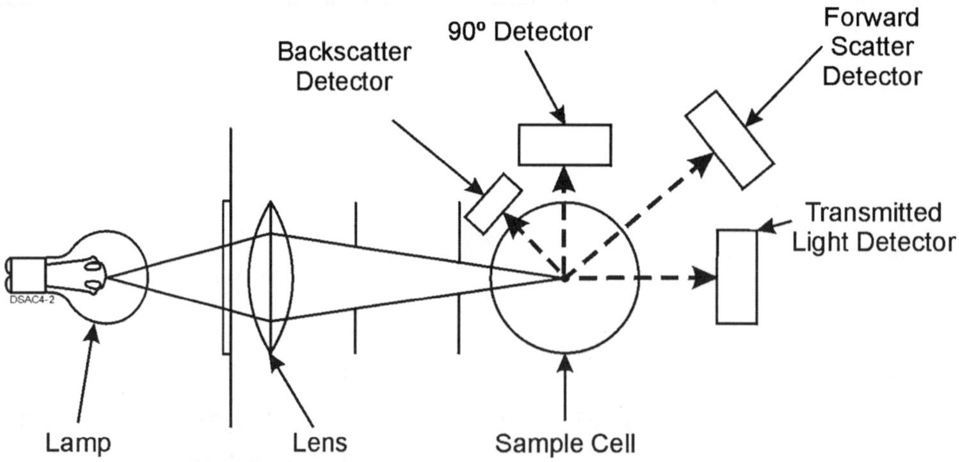

FIGURE 4-2. Optical design of a Hach ratio turbidimeter. (Adopted from M.J. Sadar, *Understanding Turbidity Science*, Technical Information Series Booklet No. 11. Hach Company, ©1996. Used with permission.)

Options available for the COTS devices include an alarm relay, a debubbling or bubble-rejecting chamber, calibration with a turbid glass cube, and automatic color compensation. Costs for sensors range from about $100 to $3,000 for sensor/analyzer units, depending on their complexity and the type of technology employed. The sensors presented are limited to

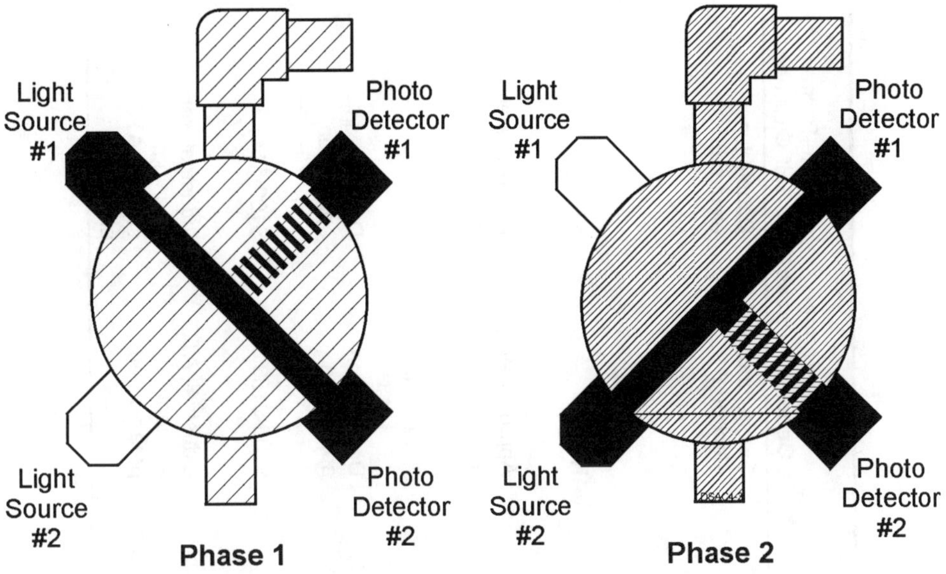

FIGURE 4-3. Configuration of the modulated four-beam method. (Copyright © GLI International, Inc. Adopted from OMEGA Engineering, Inc., *The pH and Conductivity Handbook*, 1995. Reproduced courtesy of OMEGA Engineering, Inc.)

those that monitor turbidity in NTUs, which is the standard unit of measure in the drinking water industry.

4.3 pH Sensors

Measurements of pH using electrodes have changed little in the last 50 to 60 years. A sensing device generally consists of a pH-sensing electrode, a signal amplifier, and a reference electrode. The pH electrode acts as a transducer that generates and transmits different levels of voltage based on the concentrations of free hydrogen ions in water. The amplifier permits the production of a signal amplitude large enough for measurements to be made. The isolated reference electrode produces a baseline signal that is compared to the response from the active pH electrode, making it possible to generate a pH unit measurement.

Because pH is one of the most often measured parameters in industry, pH sensors are readily available from various vendors and suppliers. Sensor options may include single- or double-junction probes, self-cleaning features, combination probes (that include pH along with other water quality parameters), glass or epoxy construction, sealed or refillable reference electrodes, and automatic temperature compensation (ATC). When reviewing literature from the manufacturers, special attention was paid to sensors with rugged design (intended for industrial use) or with reportedly minimal calibration and maintenance requirements.

Sensors from eight suppliers were reviewed for pH monitoring (Table 4-2). The cost of the sensors ranges from about $70 to $900. Because of the wide availability, only representative sensors were selected and presented for their amenability and long-term performance.

TABLE 4-2. Features of COTS pH Sensors

Manufacturer/ Model	GLI International LCP Sensor/672P Analyzer	GLI International 6070PO	Orion 91-07	Solomat 407pHX	Cole-Parmer General-Purpose	Cole-Parmer Self-Cleaning
Technology	Differential electrode technique	Differential electrode technique	Triode™ 3-in-1 pH ATC probe	Combination electrode	Single-junction electrode	Double-junction electrode
Range/ Sensitivity	0-14 pH; −500 to 500 mV and −10.0 to 100.0°C	0-14 pH	0-14 pH	0-14 pH	0-13 pH	NR
Accuracy	Nonlinearity is 0.05% of span	NR	0.05-0.1 pH; ±1°C for temperature	± 0.2 pH	NR	NR
External Dimensions	Sensor is 7.1" L with 1-1/2" NPT threading; analyzer is 5.57" x 5.67" x 5.46"	Sensor/mounting assembly are 27" L; 2" NPT threading	110 mm L x 12 mm diameter	4.13" L x 0.47" diameter	Approximately 4" in length	Approximately 4" in length
Calibration/ Maintenance Requirements	Automated one- or two-point calibration	NR	Little or no maintenance needed	Can be refilled but usually are disposed of	Depends on application	Depends on application
Cost	$545 for sensor; $885 for analyzer	$675 for sensor	$134-$150	$199	$105-$150	$100-$141
Comments	Built-in preamplifier; ATC	ATC; integral preamplifier or transmitter	Epoxy body; ATC	Simultaneous pH and temperature measurement/compensation	Industrial; ATC element; preamplified	Industrial; self-cleaning; flat surface; ATC element

TABLE 4-2. (continued)

Manufacturer/ Model	Cole-Parmer In-Line	OMEGA 5460	Global Water WQ201	GLI International 6622/2622	Innovative Sensors Model 10	Rosemount 399
Technology	Combination electrode	Double-junction combination electrode	Single-junction combination electrode	Combination electrode	Single junction	Self-cleaning, disposable combination electrode
Range/ Sensitivity	0-13 pH	0-14 pH	0-14 pH	0-14 pH	NR	0-14 pH
Accuracy	NR	0.01 pH	1% of full scale	Stability is <0.05 pH unit/24 hr	NR	NR
External Dimensions	5" L x 15/16" diameter	NR	0.8" x 8"	5.9" L x 0.875" diameter	5" in length; 0.5" NPTM	5.67" L x 1" diameter
Calibration/ Maintenance Requirements	Depends on application	Have gone 6 months without calibration	6 months to 1 yr	NR	Recommended on a monthly basis	NR
Cost	$103-$124	$105	$440 for sensor; $285 for display	$135	$70 probe; $75 preamplifier	$291
Comments	Disposable; industrial; rugged design; single or double junction	Self-cleaning; flat surface; optional ATC element	Rugged, industrial-grade, temperature compensation	Available with temperature compensation	No ATC	Integral preamplifier; ATC

ATC = automatic temperature compensation.
LCP = liquid crystal polymer.
NPT = normal pipe thread.
NPTM = normal pipe thread, male.
NR = not reported.

4.4 Chemical Dose/Residual Chlorine Monitors

Because chlorination is generally the preferred method of disinfection at small treatment systems, the review for chemical dose/residual chlorine monitoring devices was limited to that analyte. The review of manufacturers' literature for devices to monitor free chlorine resulted in the identification of four COTS technologies, including the N,N-diethyl-*p*-phenylenediamine (DPD) colorimetric method, the iodometric method, polarographic membrane sensors, and amperometric electrodes. Because chlorination has such a widespread industrial application, chlorine monitors are commercially available as simple, independent sensing devices or as complex control and alarm systems.

The DPD colorimetric method is based on American Society for Testing and Materials (ASTM) Standard Method 408.E in which DPD is oxidized by chlorine to form two oxidation products, with one being darker than the other. The level of color produced reflects the amount of free or total chlorine present in the sample, and is measured by either a colorimeter or a spectrophotometer. Some of the monitors identified require manual addition of reagents, whereas others are fully automated, including calibration. Fully automated systems may be operated unattended for up to 30 days.

The iodometric method involves adding potassium iodide to a water sample to react with available chlorine to form iodine. The amount of free iodine generated is monitored and correlated to the amount of chlorine present. This method, however, cannot distinguish between free and combined chlorine and, therefore, may be used only for monitoring total chlorine. One automatic monitoring unit was identified for this online application. In addition, it is possible to obtain a standalone sensor that measures iodine production, but this type generally is used under laboratory conditions following manual reagent addition.

The polarographic membrane sensor technology consists of a pair of electrodes that monitor free chlorine (Figure 4-4). The electrodes are immersed in a conductive electrolyte and isolated from the sample by a chlorine-permeable membrane. Free chlorine diffuses through the membrane and is reduced to chloride on the surface of the electrodes, generating a flow of electrons between the two electrodes. The current generated is proportional to the concentration of free chlorine. An automatic system that uses a polarographic membrane was identified during the literature search.

The amperometric electrodes consist of two combination probes that use a platinum cathode and a silver anode electrode to amperometrically measure free residual chlorine along with pH and temperature (Figure 4-5). Within these electrodes, an electrochemical reaction occurs based on the available chlorine concentration, generating a proportional current. These monitors rely on the proper functioning of each of the individual sensors for valid measurements to take place.

Literature was obtained for chlorine monitors from six manufacturers. Features of these residual chlorine monitors are presented in Table 4-3. Options among the various devices include ATC, pH correction, and control and alarm functions. Costs range from about $400 for standalone sensors to $10,000 for the highest level of sophistication, including complete sampling and analysis automation combined with complementary pH and temperature

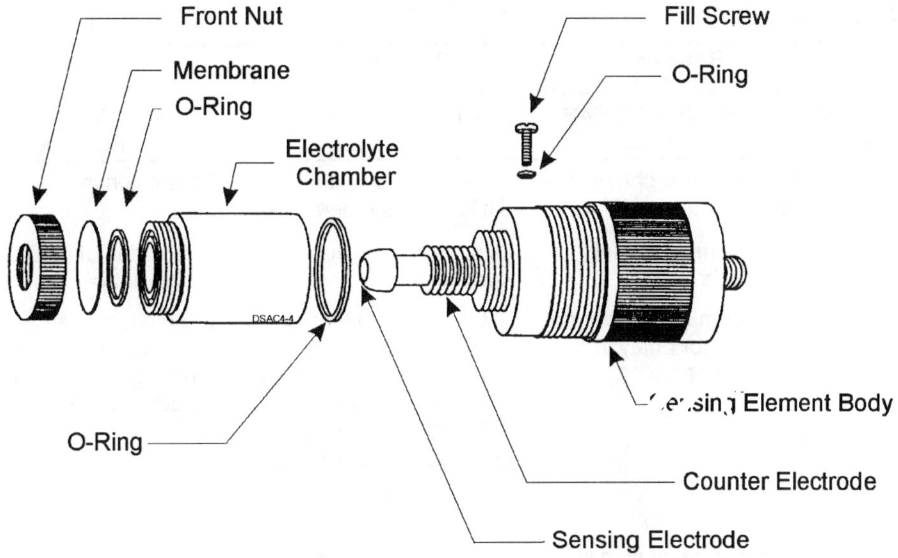

FIGURE 4-4. Components of a polarographic membrane electrode. (Adopted from ATI, Inc., *Model A 15/62 Free Chlorine Monitor, Operation & Maintenance Manual*, 1997. Used with permission.)

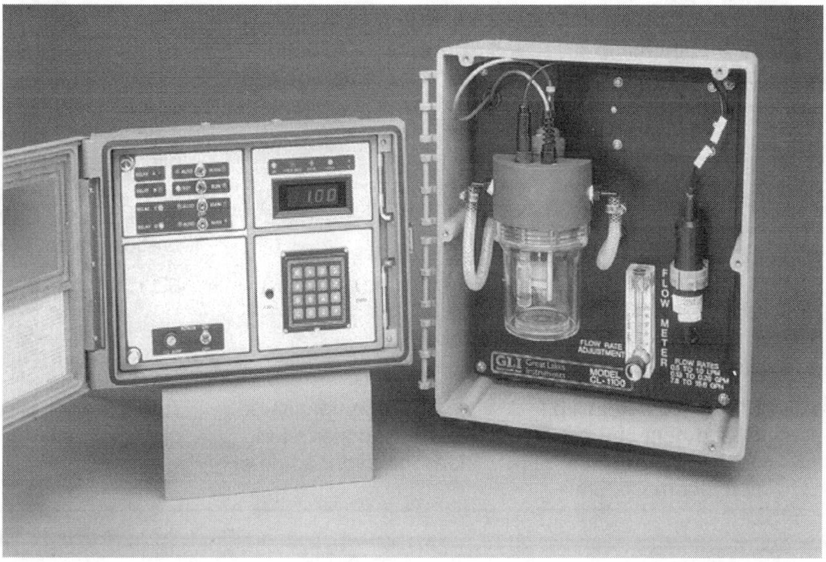

FIGURE 4-5. Residual chlorine monitoring system. (Manufactured by GLI International, Model 95 CL Analyzer and Model CL-1100 Sampling Assembly, Data Sheet 95CL/1197. Used with permission.)

compensation and control options. The cost for a standard online process control instrument with sampling assembly and analyzer is approximately $2,000 to $10,000.

TABLE 4-3. Features of COTS Chemical Dose/Residual Chlorine Monitors

Manufacturer/ Model	Rosemount Model 499A CL Sensor with Model 1054B Analyzer	GLI International Model 95 CL Residual Chlorine Monitoring System	Hanna Instruments PCA 300
Analyte	Free residual chlorine, pH, temperature	Free residual chlorine, HOCl, pH, temperature	Free chlorine
Technology	Amperometric, dual electrode	Amperometric, dual electrode	Adaptation of DPD 330.5 (colorimetric)
Range/ Sensitivity	0-20 ppm (mg/L) chlorine, 0-14 pH for analyzer; 32-122°F (0-50°C)	0.00-5.00 ppm for free residual chlorine, 0.00-5.00 ppm for HOCl, 0.00-14.00 for pH, 0.0-100.0°C for temperature; sensitivity is 0.01 pH, 0.05 ppm, and 0.1°C	0-5 mg/L range; 0.01 mg/L resolution; 0.05 mg/L minimum detectable level
Accuracy	Analyzer resolution is 0.001 ppm free residual chlorine (as Cl_2) and 0.01 pH unit	0.1 pH, 0.1 ppm, and 0.1°C	±5% of reading or ±0.05 mg/L, whichever is greater
External Dimensions	Sensor is 5.63" long and 1" in diameter; analyzer is 144 x 144 x 192 mm (5.7 x 5.7 x 7.6")	Analyzer is 12.5 x 9.7 x 6.57"; sampling assembly is 14.5 x 13.75 x 7.63"	12.5 x 10.5 x 6.25" (318 x 267 x 159 mm)
Calibration/ Maintenance Requirements	Two point buffer pH calibration standard for automatic pH correction	Replaceable chlorine sensor membrane cap; buffers and reagents not required	Reagents include indicator, buffer, and DPD compound (life is 50 days based on 10-min sampling rate); no calibration necessary, but can be done with the built-in Calibration Check System
Cost	$1,497 for microprocessor analyzer; $663 for chlorine sensor; $291 for disposable pH sensor	$1,550 for analyzer; $2,565 for sampling assembly	$2,150
Comments	Should be used with Rosemount 399 pH sensor; ATC; pH correction; alarm system	ATC; pH-compensated; control and alarm functions for chlorine and pH	Two level setpoints for process control

4.5 Temperature Monitors

Temperature is a process parameter measured extensively in both manufacturing and production operations. The most used methods for monitoring temperature involve either thermocouples or thermistors. Thermocouples operate on the principle that two dissimilar metals, connected at one end to form a junction, produce a small thermoelectric voltage when heated.

TABLE 4-3. (continued)

Manufacturer/ Model	ATI Series A15	Hach CL17	Orion Electrode	Orion Model 1770
Analyte	Free chlorine	Free or total residual chlorine	Total chlorine	Total residual chlorine
Technology	Polarographic membrane sensor	DPD colorimetric method	Based on ASTM iodometric method	Iodometric method
Range/ Sensitivity	0-2.000, 0-20.00 ppm, and 0-200.0 ppm; sensitivity is 0.001 ppm above 0.020 ppm	0-5 mg/L range (free or total residual chlorine); 0.01 mg/L resolution; minimum detection 0.05 mg/L	0.01-20 ppm	0.001-10 ppm; detection limit of 0.001 ppm
Accuracy	Linearity is 0.5%	Better than ±5% of reading or ±0.05 mg/L, whichever is greater	±2% of value between 0.05 and 50 ppm	Error less than ±10% of reading
External Dimensions	Probe is 5.5" L by 2.5" diameter; monitor is 3.5 x 3.5 x 5.5"; flow cell is 7.5 x 4 x 3"	302 mm W x 371 mm H x 175 mm D	13.9 cm L x 12 mm diameter	60"H, 25" W, 13.5" D
Calibration/ Maintenance Requirements	Does not require chemical addition	Low maintenance, 473 mL each of indicator and buffer per month; pump module tubing to be replaced at 3- to 6-month intervals	Requires iodide source	Two-point, chemical calibration; reagent consumption is 5 gal/month
Cost	$2,525 for free chlorine; $2,350 for chloramine configuration	$2,375	$389	$9,825
Comments	Can monitor chloramine if has separate sensor; control and alarm relays; ATC; needs flow cell to monitor free chlorine	Changes from free to total chlorine by adjusting reagents; warning and shutdown alarms	NR	Wall mounting

ASTM = American Society for Testing and Materials.
HOCl = hypochlorous acid.
NR = not reported.
ppm = part(s) per million.

This thermoelectric voltage is processed by sensor electronics to produce a temperature measurement. Thermocouples operate over wide temperature ranges. They are readily available to provide temperature measurements by direct insertion into gaseous, aqueous, or solid matrices.

Thermistors are manufactured from metal oxides and exhibit a predictable electrical resistance relative to the temperature to which they are exposed. A known voltage is applied to the thermistor, and based on the temperature of the matrix being monitored, a known voltage output is produced. Thermistors are configured in a simple wire design that is coated with a material permitting thermal transfer while protecting the sensor from corrosion, abrasion, or other environmental conditions that may degrade the device. Thermistors are installed the same way as thermocouples.

A noncontact temperature-measuring device identified during the literature search is the infrared pyrometer. This type of temperature sensor utilizes the principle of emissivity to determine surface temperatures. Emissivity is the ratio of energy radiated from an object at a given temperature compared to energy from a blackbody or perfect radiator at the same temperature. This type of temperature sensor generally is used to monitor moving objects or machinery that produce excessive vibrations or that are located in such a hostile environment that use of contact measurements is not possible.

The literature search also identified many suppliers of conventional glass bulb thermometers, which typically are used under controlled laboratory conditions. Glass thermometers could be used during calibration activities of RTS temperature measuring devices, but would not lend themselves to remote field applications. As such, bulb thermometers were not included as an option for future evaluation.

The literature search for temperature sensors resulted in four suppliers/manufacturers (see Table 4-4). In every case, thermocouples and thermistors are the sensors that offer the most options for industrial use. Costs range from \$30 to about \$100 for the probe with additional costs of about \$150 for the electronics that supply power and provide the readout. Complete IR pyrometers generally cost from \$500 to several thousand dollars. Accuracy of ±1°C is typical for thermocouples and thermistors, and these instruments require little or no maintenance.

4.6 Total Organic Carbon Monitors

The search for TOC-monitoring devices resulted in the identification of two instruments (Table 4-5). These instruments potentially can be integrated into online operations. Presently, TOC measurements are performed under laboratory conditions.

The first TOC monitor reviewed uses selective membrane conductivity detection and strong ultraviolet (UV)/persulfate oxidation. The system diagram for this type of analyzer is presented in Figure 4-6. The first step involves the UV oxidation reaction using persulfate to convert organic compounds to CO_2. A gas-permeable membrane, which separates the sample stream from a deionized (DI) water-containing chamber, permits the migration of CO_2 and other gaseous by-products generated, but rejects other compounds (see Figure 4-7). The conductivity of the water is monitored as the CO_2 dissolves in the DI water. The change in conductivity is proportional to the TOC level in the sample that produces the CO_2. Consumable reagents used by this monitor include phosphoric acid and persulfate.

The second TOC analyzer employs thermal oxidation of organic carbons in the sample to produce a measurable CO_2 stream. Figure 4-8 presents a schematic for such a system. In the

TABLE 4-4. Features of COTS Temperature Sensors

Manufacturer/ Model	Barnant 685-1771	Davis Instruments ST21M Sensor	Global Water WQ101	OMEGA GJQIN-18-12	OMEGA ON-970-44004	OMEGA RTD-810	OMEGA OS-614
Technology	Bimetallic conduit, iron-constantan	Type T thermocouple	Bimetallic conduit	Type J thermocouple	Thermistor	Resistance temperature device	Infrared pyrometer
Range/ Sensitivity	−31 to 200°F	−75 to 200°F	−50 to 50°C	−250 to 750°C	−40 to 100°C, 2252 ohms	−200 to 750°C, 100 ohms	0 to 250°F
Accuracy	±2°F or ±0.3% of reading	±1.5°F	±0.1°C	±0.4% full scale Resolution 1°C	±0.3% full scale Resolution to 0.1°C	± 0.3% full scale Resolution to 0.1°C	±1°F
External Dimensions	0.118" D x 4" L Install in ⅛" NPT	<0.25 lb	0.5 lb 0.625" D x 8" L	Probe 0.125" D x 12" L; conditioning module 5.7" x 3" x 1.5"	Probe 0.125" D x 5" L; conditioning module 5.7" x 3" x 1.5"	Probe 0.16" D x 4.5" L; conditioning module 5.7" x 3" x 1.5"	Probe 1.3" W x 1.3" H x 6.2" D; display 3.2" W x 1.3" H x 6.2" D
Calibration/ Maintenance Requirements	Factory-calibrated/ field confirmation	Field calibration to determine linearity of response to output from thermocouple	Factory-calibrated/field confirmation	Factory-calibrated/field confirmation	Factory-calibrated/ field confirmation	Factory-calibrated/field confirmation	Factory-calibrated/field confirmation
Cost	$30 for sensor; $150 for transmitter	ST21M, $75; transmitter, $150	$200	Probe, $30; module, $142	Probe, $55; module, $101	Probe, $79; module, $142	$650
Comments	Industrial-grace sensor, direct immersion	Surface patch thermocouple, nonintrusive, easily installed on existing system	Industrial-grade sensor, direct insertion or immersion into water	Direct immersion or placed in a thermowell	Industrial-grade thermistor, direct insertion into water through pipe via ⅛" NPT fitting	Industrial-grade sensor, direct insertion through pipe via ⅛" NPT	Noncontact measurements

TABLE 4-5. Features of COTS Total Organic Carbon Monitors

Manufacturer/ Model	Sievers 800 TOC Analyzer	Maihak TOCOR 20
Technology	Selective membrane/conductivity detection cell and strong UV/persulfate oxidation	Thermal oxidation of organic carbon and subsequent detection of CO_2
Range/ Sensitivity	0.05- to 50,000 ppb in hot or cold water	0-50 mg/L; 0-100 mg/L; 0-200 mg/L; 0-500 mg/L; and 0-1,000 mg/L
Accuracy	NR	NR
External Dimensions	6" H x 19.5" W x 13.5" D	1200 x 600 x 500 mm
Calibration/ Maintenance Requirements	Quarterly replacement of reagent modules; semiannual UV lamp and annual pump tubing replacement; excellent stability for up to 1 yr; easy-to-verify calibration using standard solutions	NR
Cost	$19,800 for analyzer; $150/box for acid and oxidizer	Analog unit, $19,000
Comments	Easy data transfer to PC, self-diagnostics, remote alarms and controls	Optional dilution system and backflush filter

NR = not reported.
PC = personal computer.
ppb = part(s) per billion.

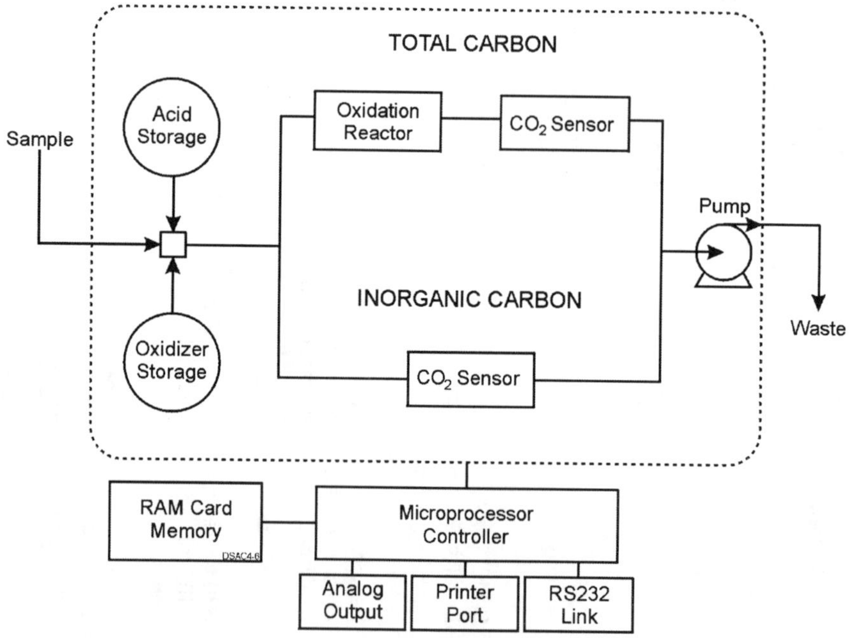

FIGURE 4-6. System diagram of a TOC analyzer using selective membrane conductivity detection and strong UV/persulfate oxidation. (Adopted from Sievers Instruments, Inc., Model 800 TOC system diagram. Used with permission.)

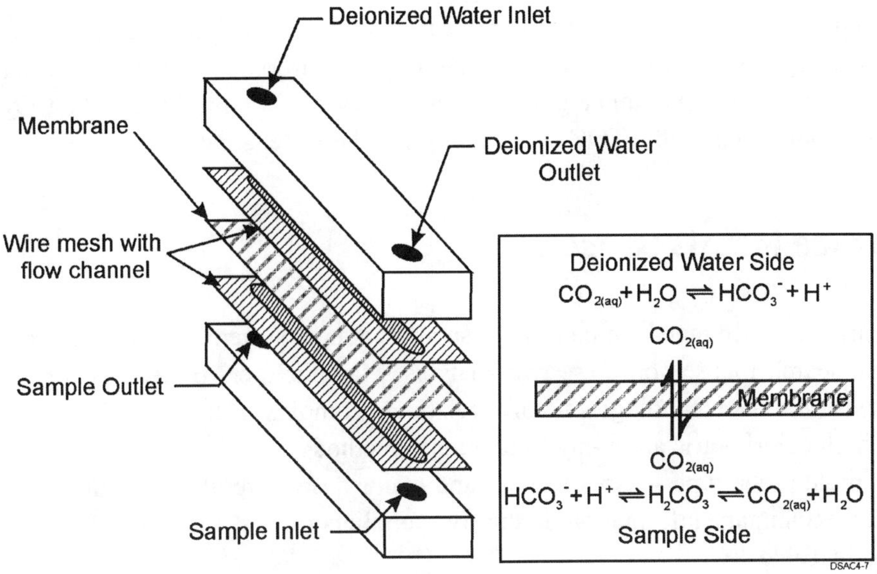

FIGURE 4-7. Schematic of a membrane module and equilibrium reactions. (Adopted from R. Godec et al., "New Technology for TOC Analysis in Water." *Ultrapure Water Journal*, ©1992. Used with permission from Tall Oaks Publishing, Inc.

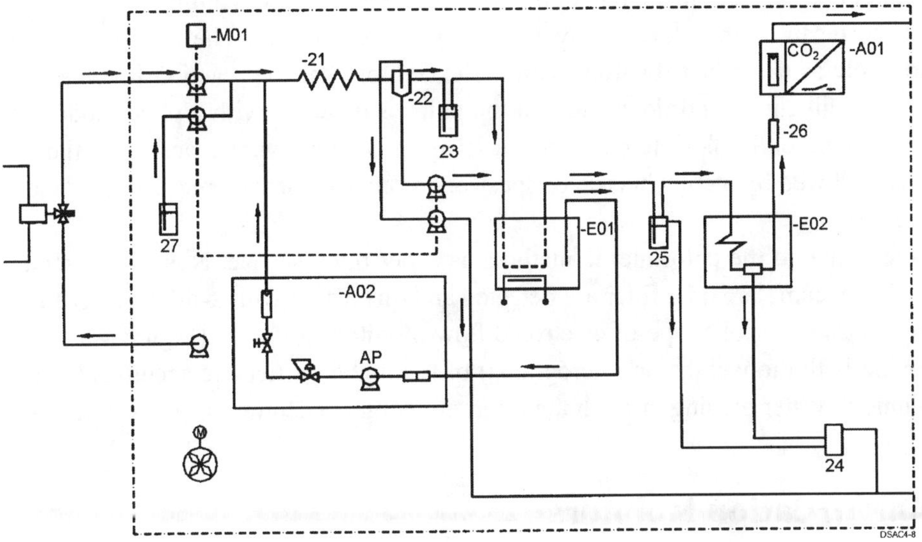

- A01	CO_2 Analyzer	- 22	Phase Separator
- A02	Control Unit	23	CO_2 Absorber
- M01	Peristaltic Hose Pump	24	Syphon
- E01	Thermal Reactor	25	Gas Washer
- E02	Sample Gas Cooler	- 26	Sacrificial Filter
- 21	Stripper	27	Acid Supply

FIGURE 4-8. Schematic of a TOC analyzer using thermal oxidation to produce CO_2 for measurements. (Adopted from Maihak data sheet #80.500 USA/97.10. Used with permission from Maihak Aktiengesellschaft.)

initial step, any inorganic carbon is removed by a cascade stripper. A thermal zone then achieves catalytic oxidation of the remaining organic carbons. A condensate separator removes excess water vapor, and a gas analyzer titrates the CO_2 produced. The CO_2 concentration is proportional to the TOC concentration in the water sample.

4.7 Trace Metals Sensors

The primary trace metals of concern in small treatment systems are iron and manganese. Thus, the search focused on these two analytes. Manufacturers of sensors for these trace metals exist, but the technologies involve relatively complex instruments. The technologies include both colorimetric and x-ray fluorescence methods.

No simple probe-type sensors for iron and manganese were identified during the literature search. Summary information on the five suppliers of iron and manganese monitors is provided in Table 4-6.

The colorimetric method is available for both iron and manganese. An iron monitor that was identified (Table 4-6) meters the water sample with a reagent, 2,4,6-tripyridyl-*s*-triazine (TPTZ), under acidic conditions, to produce a color proportional to the iron concentration. The processed sample is pumped automatically into an optical cell where light transmission is measured at 590 nm with a photodiode detector. Transmission/absorbence calibration curves permit the quantitation of the iron present in the sample.

Manganese can be monitored by two colorimetric techniques. The first operates on the same principle as described for iron, with the exception that the reaction involves formaldoxime under alkaline conditions and that the color is detected with a photodiode detector at 450 nm. A second colorimetric manganese monitor employs a reaction of the sample with DPD followed by a colorimetric or spectrophotometric measurement to produce quantitative results.

An alternative to the colorimetric methods is x-ray fluorescence. Nonsoluble metal particulates that accumulate on a filter are detected and quantified with x-ray fluorescence. The instrument uses a Coriolis type of electronic flow-monitoring device (Figure 4-9). The raw data provide both the weight (in micrograms) of the metal particulate accumulated and the total volume of water passing through the filter, allowing concentrations to be calculated.

4.8 Anion/Cation Monitors

Two anions, nitrate and fluoride, were specified as being of interest for small treatment system monitoring, so the review was limited to these two analytes. Search results indicate that COTS ion-selective electrodes (ISEs) are available for nitrate and fluoride.

The technologies identified involve reading a millivolt or milliamp signal from ISEs. An ion-selective electrode consists of a sensing and a reference electrode. The reference electrode provides a zero-response condition, whereas the sensing electrode provides an electrochemical current produced due to the presence of an anion of interest. The potential

TABLE 4-6. Features of COTS Iron and Manganese Sensors

Manufacturer/ Model	Tytronics® Manganese	Tytronics® Iron FPA 806	Rosemount Chempure™	Capital Controls Aztec® M100	Capital Controls Aztec® FE100	Hach 31800	Detora On-Line Metals
Analyte	Manganese	Iron	Iron	Manganese	Iron	Manganese	Metals (atomic numbers 13-92)
Technology	Online colorimetric method	Online colorimetric method	Online colorimetric method	Formaldoxime colorimetric method	TPTZ colorimetric method	DPD pump-colorimeter	X-ray fluorescence
Range/ Sensitivity	0-200 ppb	0-500 ppb; 0-1 ppm; 0-10 ppm; 0-100 ppm	0-300 ppb; 0-1 ppm; 0-10 ppm	20-500 ppb Mn^{2+}	0.020-2.00 mg/L	0-0.2 mg/L	Sub-ppb to percent; detection limit of 0.007-0.01 ppb
Accuracy	NR	NR	NR	5% at full range	±5%	±5% of range	NR
External Dimensions	NR	NR	44 x 16 x 18"	NR	NR	16.25" W x 20.25" H x 11.7" D	25" x 26" footprint
Calibration/ Maintenance Requirements	pH buffer and formaldoxime reagents; three calibration methods	Reagents are ferrozine/ hydroxyl-amine and acetate buffer	15 min monthly for reagent and tubing replace-ment; automatic calibration	Two-point autocali-bration; minimum of 20 days without reagent replacement	Two-point autocalibration; self-cleaning optics; minimum of 20 days without reagent replacement	Weekly standardiza-tion and buffer replenishment; replace tubing every 3 to 6 months	No reagents; low maintenance; no waste; calibration using prepared standards; replace PVDF filters
Cost	$8,950 for single stream	$8,950 for single stream	$13,778 for analyzer; $2,255 for 1-yr reagent supply	$18,000; $380/month for reagents	$18,000; $186/month for reagents	$4,595	$125,000 for particulate Fe and Mn; $187,000 for dissolved
Comments	Batch sampling minimizes reagent consumption; alarm function	Self-diagnostics	Self-diagnostics	Self-cleaning optics; self-diagnostics and alarm system; rugged industrial design	Self-cleaning optics; self-diagnostics and alarm system	Alarm system; chlo-rine can be interfer-ence; model will be discontinued soon due to system performance	100 mCi Cm-244 x-ray source

NR = not reported.
PVDF= polyvinylidene difluoride.

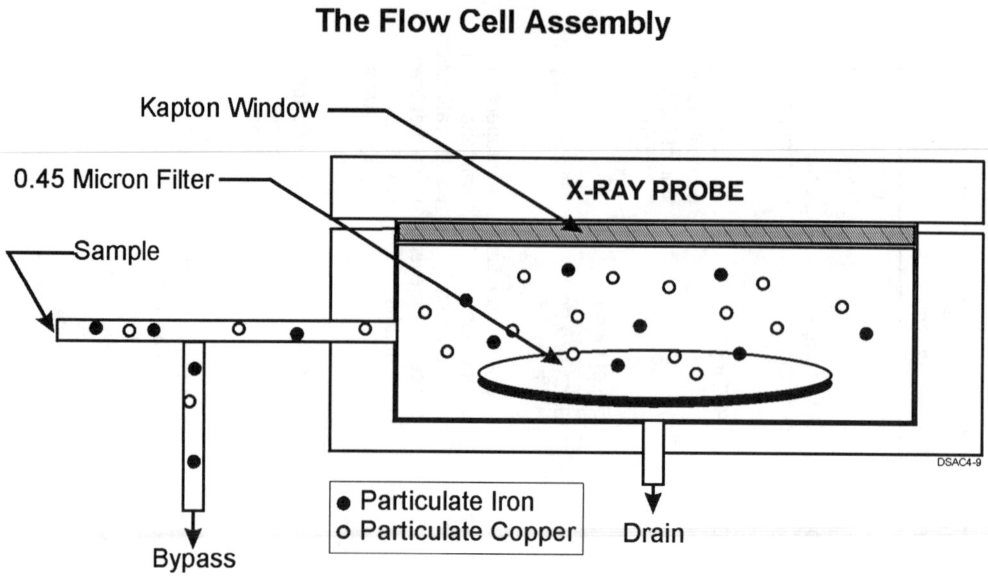

FIGURE 4-9. X-ray fluorescence monitoring for iron. (Adopted from Detora Analytical, Inc., *OCPM: On-Line Metals Monitoring System*. Used with permission.)

difference measured between the two electrodes, when correlated against a known concentration response curve, permits quantitative analyses.

Process monitors based on ISE probes are available. These instruments provide the required buffers and have multiple standard additions and automatic electrode-refreshing features.

Manufacturers' literature for 10 nitrate and 7 fluoride monitors was reviewed (see Tables 4-7 and 4-8 for selected features). Many other models with slightly different features also are available commercially. Costs of fluoride electrodes range from $200 to $600, and costs of associated meters range from approximately $700 to $2,000. An online fluoride-monitoring system with alarm relays costs up to $18,000. Costs of nitrate electrodes range from approximately $200 to $700, with meters costing approximately $700 to $2,000. Process analyzers for nitrate monitoring cost from $6,000 to $18,000.

4.9 Color Monitors

Pump colorimeters that monitor distinct changes in color are readily available; however, few applications exist that are designed to monitor color in potable water. Online instruments identified provide outputs in optical density, but not in platinum-cobalt color units which is the standard unit of measurement for water.

Spectral-absorbence technology is one option for color monitoring. This technology measures the spectral absorbence of the water as it flows through a sample chamber. A specific wavelength of energy is produced by a source lamp and through optical filters, and the level of color is measured by a photometric detector (see Figure 4-10). A second monitor

TABLE 4-7. Features of COTS Nitrate Monitors

Manufacturer/ Model	Solomat 603IS	OMEGA	Cole-Parmer Nitrate	Orion 930700	Orion Ionplus
Technology	Ion-selective electrode	Ion-selective combination electrode	PVC membrane electrode	Plastic membrane half-cell electrode; direct method	Plastic membrane combination; direct method
Range/ Sensitivity	0.18-230 ppm	62,000-0.5 ppm for sensor; 0.001-1990 for meter	0.5-62,000 ppm for electrode; meter is 0-9999 for concentration and ±1999 for mV	1 to 7 x 10^{-6} M; 14,000 to 0.1 ppm as N	1 to 7 x 10^{-6} M; 14,000 to 0.1 ppm as N
Accuracy	NR	±0.5% for meter	±0.5% of reading for concentration; ±0.2 for mV	±0.5% of reading	±0.5% of reading
External Dimensions	Analyzer is 2.9"-4.4" W x 1.6"-2.6" D x 8.3" L	8.3" x 12" x 4.1" for meter	9" L x 7" W x 2.375" H for meter	5.375" L (13.5 cm)	NR
Calibration/ Maintenance Requirements	NR	Two-point automatic calibration	NR	NR	NR
Cost	$749 for electrode; $239 for reference electrode; $1,395 for display meter	$206 for PVC; $275 for glass	$340 for electrode; $795 for meter	$555 for electrode; $690 for meter	$625 for electrode; $690 for meter
Comments	BNC plug	pH range from 2.5-11; manual or automatic temperature compensation	Single- or double-junction; pH range is 2.5-11; manual or automatic temperature compensation	PVC gelled membrane and body	Membrane replacement recommended every 3 months

measures absorbence in the water stream by using side-by-side photodetectors. The instrument actually monitors color differences between the two detectors and, if it is calibrated with known "high" and "low" standards, it can be programmed to provide absolute readings.

A second color-measuring technology, which covers the spectrum of the human eye, is a potentiometric method that involves a light source and a sensor head located across from each other. This device generally is used with solids; however, it can be adapted to a liquid medium. Accuracy decreases when it is used for liquids due to interferences associated with air bubbles in the sample stream. The device has a spectral sensitivity of 400 to 700 nm.

Four suppliers of color-monitoring systems were located, and their monitors are presented in Table 4-9. Numerous other color-monitoring devices are available for online industrial applications, but discussions with the suppliers indicate that they most likely are not sensitive to or appropriate for detecting color in potable water. The costs of the absorbence systems range from $6,000 to $9,000. The costs of the potentiometric devices are about $1,900.

TABLE 4-7. (continued)

Manufacturer/ Model	Hach One Nitrate	Hach APA 6000	Corning Combination	Corning Half-Cell	Capital Controls Aztec® N1000
Technology	Combination electrode	Process analyzer with ISE	Direct; ion-exchange membrane	Direct	Ion-selective, nitrate-sensing electrode
Range/ Sensitivity	0.10-14,000 mg/L	0.1-100 mg/L	1 to 5 x 10^{-6}M; 62,000 to 0.8 ppm	1 to 5 x 10^{-6}M; 62,000 to 0.8 ppm	Automatic ranging from 0-500 mg/L NO_3
Accuracy	NR	NR	±0.5% of ISE reading	±0.5% of ISE reading	3% of reading
External Dimensions	9.4 x 0.41"	NR	5.5" L (14 cm)	5.5" L (14 cm)	21.5" x 26.5" x 8"
Calibration/ Maintenance Requirements	NR	Can set up for automatic daily calibration	NR	NR	Automatic high and low calibration with nitrate standard solutions
Cost	$315 for electrode	$6,200 with AquaTrend interface	$450 for electrode; $1,995 for meter	$435 for electrode; $1,995 for meter	$18,000
Comments	Double-junction; pH range 3-11; refillable	Four control loops	Double-junction reference probe; ATC	ATC from −30 to 130°C	Temperature-controlled sample stream and electrode; preconditioned samples

ATC = automatic temperature compensation.
BNC = bayonet N connector.
ISE = ion-selective electrode.
NR = not reported.
PVC = polyvinyl chloride.

4.10 Hardness Monitors

Technologies for measuring hardness in water range from simple ISEs to complex on-line monitoring and control systems. The ISEs reportedly can be inserted directly into the aqueous stream for continuous monitoring. These electrodes operate on the same principle as the pH and anion sensors discussed in Sections 4.3 and 4.8. Standalone ISEs for hardness are designed to respond to concentrations of calcium and/or magnesium in water. More complex monitors are available that integrate the calcium ISEs into an automated system that includes reference cells, temperature probes, and a diffusion reagent delivery process for pH adjustment.

A colorimetric method also is available to monitor hardness. The instrument uses a buffered wet-chemical colorimetric indicator. The indicator and a buffer are added to the sample to produce a color measurable by a spectrophotometric detector. Samples and reagents are automatically metered and injected by a pump/valve module.

TABLE 4-8. Features of COTS Fluoride Monitors

Manufacturer/ Model	Orion Ionplus	Orion Half-Cell	Corning Combination	Solomat 601IS	OMEGA Fluoride	Cole-Parmer Fluoride	Capital Controls Aztec® F1000
Technology	Solid-state combination electrode	Half-cell and reference electrode	Double-junction combination electrode	Ion-selective electrode	Solid-state combination electrode	Solid-state electrode	Ion-selective electrode
Range/ Sensitivity	Saturated to 0.02 ppm for sensor; 0.000-19,900 for meter	Saturated to 0.02 ppm for sensor; 0.000-19,900 for meter	1 to 5 x 10^{-6}M; 19,000-0.019 ppm	0.02-19,000 ppm	0.02 ppm to saturated for electrode; 0.001-1990 for meter; resolution to 0.001 for concentration and to 0.1 mV or 1 mV	0.1 to 260 ppm for electrode; meter is 0-9999 for concentration and ±1999 for mV	Automatic ranging from 0-10 mg/L as F^-
Accuracy	2% error/degree for sensor; ±0.5% of reading for meter	2% error/degree for sensor; ±0.5% of reading for meter	±0.5% of ISE reading	NR	±0.5% for concentration; ±0.2% ±1 digit for mV	±0.5% of reading for concentration; ±0.2 for mV	2% of reading
External Dimensions	NR	NR	5.5" L (14 cm)	Analyzer is 2.9"-4.4" W x 1.6"-2.6" D x 8.3" L	8.3" L x 12" W x 4.1" H	9" L x 7" W x 2⅜" H	21.5" x 26.5" x 8"
Calibration/ Maintenance Requirements	Recommended daily	Recommended daily	NR	NR	Two-point autocalibration	NR	Automatic high and low calibration with fluoride standard solutions
Cost	$555 for electrode; $690 for meter	$450 for half-cell fluoride electrode; $140 for reference electrode; $690 for meter	$435 for electrode; $1,995 for meter	$629 for electrode; $195 for reference electrode; $1,395 for display meter	$186 for epoxy electrode and $275 for glass electrode; $1,260 for meter	$290 for electrode; $795 for meter	$18,000; $465/month for solutions
Comments	Five-point autocalibration; should maintain sample pH of 5	Five-point autocalibration; should maintain sample pH of 5	Glass body sensor; meter has automatic temperature compensation	Bayonet N connector (BNC) plug	pH range is 5-8; manual or automatic temperature compensation	pH range is 5-8; manual or automatic temperature compensation	Self-diagnostics; alarm relays; samples with particles greater than 100 microns may require prefiltration

NR = not reported.

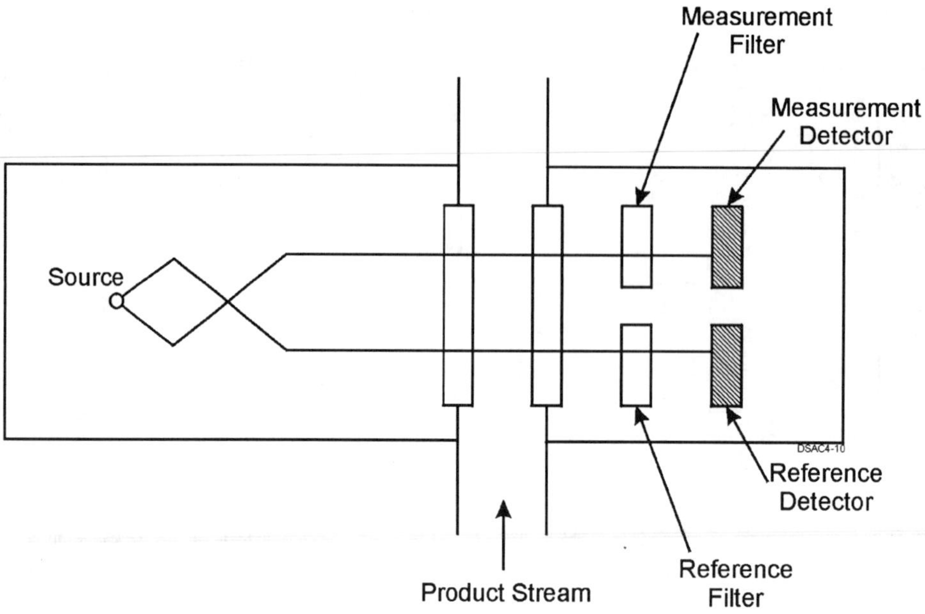

FIGURE 4-10. Color measurements using spectral absorbence technology. The ratio between the absorbence through the measurement and reference filters is directly related to the concentration of color in the product stream. (Adopted from MONITEK® In-Line Process Color Monitors, 1994. Used with permission.)

Literature on hardness-sensing devices from five manufacturers was reviewed. Features of the representative hardness monitors are presented in Table 4-10. Hardness monitors are available with ATC, automatic cleaning, online calibration, and alarm and control systems. Standalone ISEs with electronics cost from $1,000 to $2,400, whereas the fully automated COTS systems can cost up to $6,000.

4.11 Total Dissolved Solids/Conductivity Sensors

Probes to measure TDS and conductivity are readily available. Literature for nine sensors was obtained and reviewed. The sensors in Table 4-11 represent the COTS technologies for monitoring these parameters.

All of the sensors include some level of temperature compensation, and most include temperature as an output reading. TDS and conductivity readings are usually generated from a single probe. Multiparameter probes that provide TDS measurements also may include salinity, turbidity, and resistivity measurements.

Conductivity is a measure of the capacity of a solution to conduct electricity, which is the reciprocal of resistivity. The presence of dissolved solids is proportional to conductivity. An increase in dissolved solids generally increases conductivity and decreases resistivity.

Both contacting sensors and sensors without electrodes are available for monitoring conductivity and TDS. A contacting sensor consists of multiple electrodes spaced at specific

TABLE 4-9. Features of COTS Color Monitors

Manufacturer/ Model	Wedgewood AF20 Sensor/ 680 Analyzer Transmitter	Tytronics® Color Analyzer	MONITEK® CS6	Component Engineering
Technology	Spectral absorbence	Direct spectro-photometric	Sensor detects incan-descent light; silicon photodetectors measure absorbence	Potentiometer
Range/ Sensitivity	Five fixed selectable ranges or one variable range; 400-1,900 nm	450 nm and 810 nm	Full-scale ranges for transmitter are 0-100%, 0.01-5.0 absorbence units (A.U.), and other ranges available; sensor sensitivity is ±0.1% of span	Available for various target distances and sizes; spectral sensitivity of 400-700 nm
Accuracy	2% over full range	NR	Resolution is 0.02% of full scale; linearity is 0.1%	Depends on what is being monitored (i.e., if it is liquid, moving, or con-tains air bubbles)
External Dimensions	Analyzer is 5.6 x 5.06 x 6.69"	NR	Sensor mountings available for 0.5" or 1" NPT; transmitter is 14.7 x 12.3 x 6.7"	Sensor has 18 mm diameter; can be reduced to 6-8 mm with fiber optics on end
Calibration/ Maintenance Requirements	Incandescent light has 10,000- to 25,000-hr lifespan	Automatic; weekly	Coating can prevent scaling on window; change incandescent light bulb as needed	NR
Cost	$6,000 to $7,000	$8,950 for single stream	~$9,000 (low-range CS8 sensor costs ~$12,000)	~$1,900 for light source and sensor with fiber optics; ~$1,700 without fiber optics
Comments	Temperature com-pensation from 0 to 25°C; output in optical density units must be converted to color units	Batch-sampling technique; alarm functions; designed for potable water applications	Calibration with known highs and lows gives absolute readings	Optical fiber components

NR = not reported.

distances (Figure 4-11). The distance between any two electrodes creates a cell constant. Variations in the cell constant allow for the measurement of conductivity in different ranges. Sensors without electrodes introduce an alternating current into a closed loop of solution (Figure 4-12). Conductivity is determined by the magnitude of the current.

Another technology has been developed specifically for wash applications that incorporates two tin-over-nickel-plated stainless steel probes. The sensor is designed for rugged use with little or no maintenance and is low in cost. There were no reports of this sensor

TABLE 4-10. Features of COTS Hardness Monitors

Manufacturer/ Model	Solomat 635IS	Orion 1820	Hach 1220	Cole-Parmer	OMEGA	Hach 6000
Technology	Ion-selective electrode	Ion-selective electrode; diffusion reagent system	Buffered wet-chemical colorimetric indicator	PVC membrane combination electrode	Polymer membrane combination electrode	Colorimetric, automated EDTA
Range/ Sensitivity	0.4-40,000 ppm	Up to 350 ppm	0-100 mg/L	0.4 to 40,000 ppm as Ca^{2+} for electrode; meter is 0-9999 for concentration and ±1999 for mV	0.4 to 40,000 ppm for electrode; 0.001-1990 for meter; resolution to 0.001 for concentration and to 0.1 mV or 1 mV	50 to 10,000 μg/L
Accuracy	NR	±10% of reading or ±10 ppb as $CaCO_3$, whichever is greater	NR	±0.5% of reading for concentration; ±0.2 for mV	±0.5% for concentration; ±0.2% ± 1 digit for mV	NR
External Dimensions	Analyzer is 2.9"-4.4" W x 1.6"-2.6" D x 8.3" L	Analyzer is 26" x 17" x 11"	14 x 12 x 6.25'	9" L x 7" W x 2.375" H	8.3" L x 12" W x 4.1" H	NR
Calibration/ Maintenance Requirements	NR	Requires 20 min of maintenance/month; calibration with double known addition (DKA) technique; calibrate at least monthly	Consumption is 1 L each of indicator and buffer every 2 months	NR	Two-point autocalibration	Automatic calibration
Cost	$829 for electrode; $195 for reference electrode; $1,395 for display meter	$4,245 for analyzer; $1,050/yr for consumables	$1,500	$340 for electrode; $795 for meter	$220 for PVC electrode; $275 for glass electrode; $1,260 for meter	$5,800
Comments	Bayonet N connector (BNC) plug	Optional alarm module	Alarm system	Interferences are Cu^{+2}, Zn^{+2}, Ni^{+2}, and Fe^{+2}; manual or automatic temperature compensation	Interferences include Cu^{+2}, Zn^{+2}, Ni^{+2}, and Fe^{+2}; manual or automatic temperature compensation	Automatically primed, cleaned, and calibrated

NR = not reported.

TABLE 4-11. Features of COTS TDS/Conductivity Sensors

Manufacturer/ Model	Orion 142	Honeywell	Hanna Instruments HI 9635	Orion 124	OMEGA CDH-287-KIT
Analyte	Conductivity, TDS, salinity, and temperature	Conductivity, turbidity, and temperature	Conductivity, TDS, and temperature	Conductivity, TDS, temperature	Conductivity, TDS, resistivity, salinity
Technology	NR	Two tin-over-nickel-plated 303 stainless steel (SS) probes	NR	Four-electrode cell technology	NR
Range/ Sensitivity	Conductivity range is 0-500,000 μs/cm; TDS range is 0-1999 mg/L	0.00001 to 15 mS for conductivity; 0-4000 NTUs for turbidity; 68-140°F for temperature	0.0 to 150.0 ppm (mg/L), 150 to 1,500 ppm (mg/L), 1.50 to 15.00 ppt (g/L)	0-2,000 ppm TDS; 0-199,000 μS/cm in four ranges; -5 to 50°C temperature	Five ranges from 0-130 g/L for TDS; six ranges from 0-1999 mS/cm for conductivity
Accuracy	Conductivity is 0.5% ±1 digit at 15-35°C; TDS is ±2% ±1 digit at 15-35°C	See manufacturer's curve	±2% full scale excluding probe error (@20°C)	±0.5% for conductivity; ±0.1°C for temperature	± 0.5% of reading
External Dimensions	200 x 90 x 190 mm	Mounting dimensions are 1.25" diameter x 2.39"; cap is 0.71" OD x 2.75"	7.7 x 3.1 x 2.4"	Meter, 17 x 9 x 6 cm; cell, 21 mm diameter x 150 mm L	100 x 180 x 44 mm
Calibration/ Maintenance Requirements	Automated calibration	NR	Automatic one-, two-, or three-point calibration	NR	NR
Cost	$1,650	$90 without flowthrough cap; $100 with flowthrough cap	$490	Meter, $795; cell, $260	Meter with probe, $740; replacement probes, $105 to $275
Comments	Four temperature compensation algorithms	With microprocessor and software	Automatic temperature compensation from 32-140°F	Automatic temperature compensation	Automatic temperature compensation from 0-50°C

arrangement being used to measure TDS/conductivity in potable water, but its principle of operation and low cost make it an option for possible evaluation and use in an RTS system.

Conductivity/TDS probes range in price from $65 to more than $300. Associated meters and display units range from $300 to $1,700. The conductivity/turbidity sensor designed for wash applications costs approximately $100, which includes the cost for a microprocessor readout display. Other probes are available for monitoring TDS/conductivity; however, those presented in Table 4-11 are representative of COTS technologies.

TABLE 4-11. (continued)

Manufacturer/ Model	Orion Advanced Portable Multimeter	Davis Instruments 72	Hanna Instruments HI 8734	Cole-Parmer
Analyte	Conductivity, salinity, TDS, temperature	Conductivity, TDS, temperature	TDS	Conductivity, TDS
Technology	NR	Carbon electrode	NR	SS electrode
Range/ Sensitivity	0-1,999,900 μS in six ranges; −5 to 90°C for temperature	0-10,000 μS/cm for conductivity (four ranges); 0-7,000 μg/L for TDS (four ranges); −50 to 150°C for temperature	0-199.9 mg/L; 0-1999 mg/L; 0-19.99 g/L	Four ranges from 0.00-19.99 mS for conductivity; four ranges from 0.00-999 ppm for TDS; 0-80.0°C for temperature
Accuracy	0.5% ±1 digit for conductivity; ±0.5°C for temperature	±2.5%	±1% full scale	Greater of ±1% or ±1 digit for conductivity and TDS; ±0.5°C for temperature
External Dimensions	NR	Cell is 6" L x 0.5" diameter	180 L x 83 W x 40 H mm	110 mm L x 12 mm OD for cell; 3.5" W x 7.375" H x 1.75" D
Calibration/ Maintenance Requirements	NR	NR	NR	NR
Cost	$1,650 for meter; $330 for standard cell	$360 for meter; $94 for conductivity cell	Meter and probe, $298; replacement probe, $65	$395 for meter; $65 for replacement cell
Comments	Autoranging; automatic temperature compensation	Manual temperature compensation from 0-100°C	Manual temperature compensation from 0-50°C	Automatic temperature compensation from 0-50°C

NR = not reported.
ppt = part(s) per thousand.

4.12 Particle-Counting Monitors

The review of manufacturer's literature for particle-counting instruments identified two COTS instruments suitable for online monitoring. These devices normally are used to determine filtration efficiency, permitting the optimization of that water treatment process.

Both particle-counting monitors operate on the principle of light extinction/blocking. This method differs from turbidity measurements when light scattering is monitored. A solid-state laser diode and electronic photodiode permit the determination of not only particle counts but also particle size (see Figure 4-13). These capabilities could be useful as a microbial monitoring tool by looking for particles in the size range of *C. parvum* and *G. lamblia*.

Selected features of the particle-counting monitors are provided in Table 4-12.

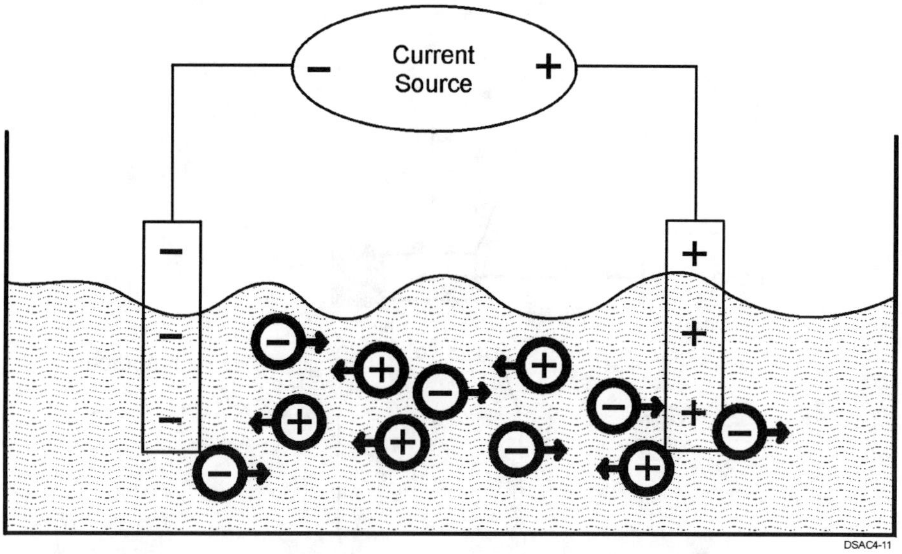

FIGURE 4-11. Contacting sensors determine the ability of a solution to conduct an electrical current. (Copyright © OMEGA Engineering, Inc., *The pH and Conductivity Handbook*, 1995. All rights reserved. Reproduced with permission.)

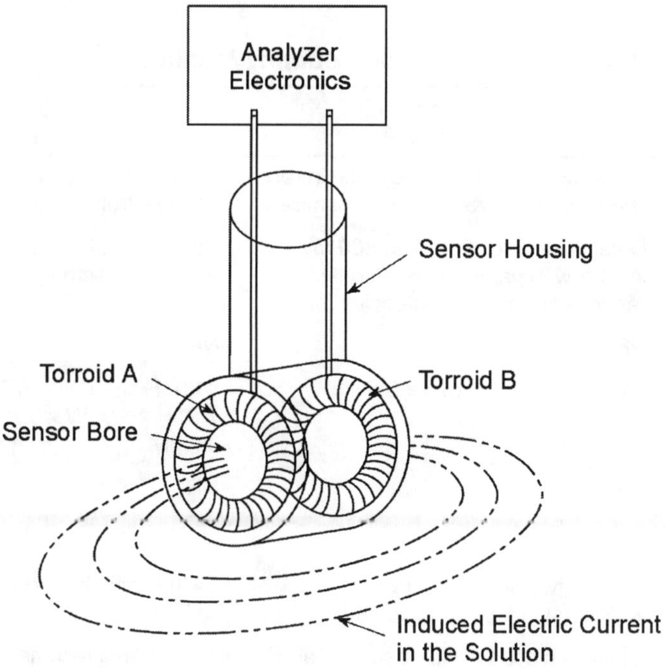

FIGURE 4-12. Operating principle of an electrodeless sensor. (Copyright © GLI International, Inc. Adopted from OMEGA Engineering, Inc., *The Flow and Level Handbook*, 1995. Reproduced courtesy of OMEGA Engineering, Inc.)

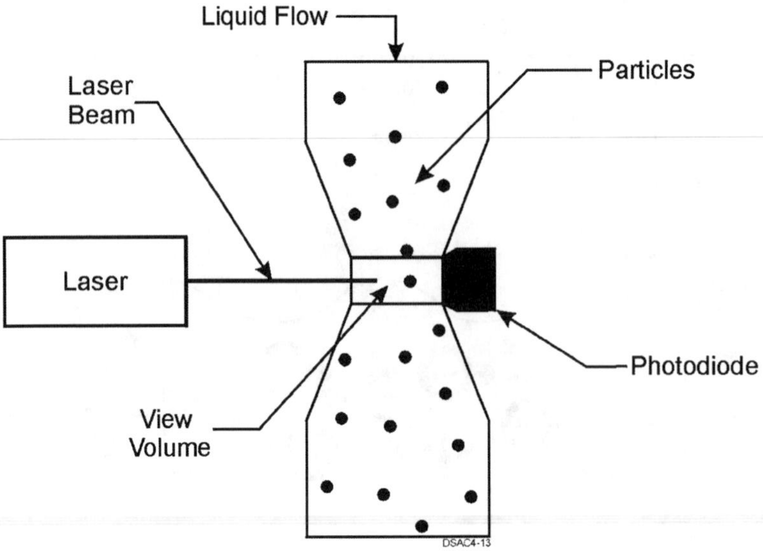

FIGURE 4-13. Particle-counting monitor that operates by measuring the amount of light blocked by suspended material in water. (Adopted from J. Hurt and S. Hjelmervik, *Light-Blocking Laser Sensors for Liquids: Product Note*. Met One, Inc., ©1991. Used with permission.)

TABLE 4-12. Features of COTS Particle-Counting Monitors

Manufacturer/ Model	Hach Model 1900	Met One PCX
Technology	Light-extinction technology with infrared laser diode to size and count particles	Light blocking (laser diode light source and electronic photodiode)
Range/ Sensitivity	Detects particles from 2 to 800 microns in size, with particle concentration ranges of 0 to 20,000 particles/mL	Up to 32 particle sizes from 2 microns up to 750 microns
Accuracy	NR	NR
External Dimensions	14 x 23 x 10.2 cm (5.5 x 9 x 4")	11.0" H x 9.0" W x 6.7" D for sensor; 7.0" H x 7.0" W x 6.7" D for power supply
Calibration/ Maintenance Requirements	NR	NR
Cost	\$9,990 (two-sensor system) \$6,325 (single-sensor system)	\$4,100 with flow controller (\$3,950 without)
Comments	Sensors placed at filter bed inlet and outlet; software program displays sample information and log reduction or percent filter efficiency; alarm capabilities; 14" of head pressure required	100 mL/min recommended sample flowrate

NR – not reported.

4.13 Multiparameter Probes

Multiparameter probes were identified during the search. Literature for several of these probes was reviewed (see Figure 4-14 for an example). Costs for multiparameter probes vary, based on the number and types of sensing units that are incorporated into the sensor head. Calibrations, consumables, and maintenance must be considered and applied to each individual device included in a multiparameter probe.

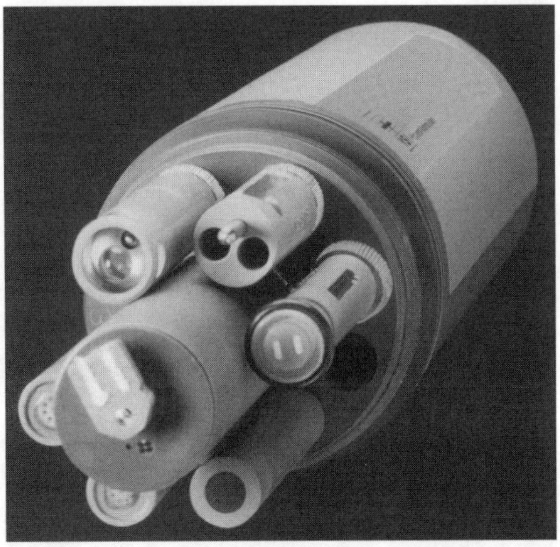

(a)

YSI 6-Series Multi-Parameter Matrix

	DO	pH	TDS	ORP	vented level	vented level	temperature	resistivity	turbidity	TDG	conductivity	specific conductance	ammonium/ammonia	nitrate	chloride	salinity
YSI 6000	√	√	√	√	√	√	√	√	√	√	√	√	√	√		√
YSI 6920	√	√	√	√	√	√	√	√	√		√	√	√	√	√	√
YSI 6820	√	√	√	√	√	√	√	√	√		√	√	√	√	√	√
YSI600XLM	√	√	√	√	√	√	√	√			√	√				√
YSI 600XL	√	√	√	√	√	√	√	√			√	√				√
YSI 600R	√	√	√				√	√			√	√				√

DSAC4-14

(b)

FIGURE 4-14. (a) An example of a multiparameter probe. (b) Various parameter configurations available in multiparameter probes. (Adopted from YSI, Inc., Environment Monitoring Systems: Advanced Technology for Water Quality Monitoring. Used with permission.)

5

COTS System Operation Monitoring and Control Devices

This chapter presents the results of the literature search for the various system operation monitoring and control devices discussed in Chapter 3. Similar to Chapter 4, spreadsheets are used to compare the various features of the COTS devices identified for each of the parameters. Individual data sheets for all COTS devices discussed in this chapter are presented in Appendix C.

5.1 Flowrate Sensors

The search for flowrate monitors resulted in a number of COTS devices, including rotameters, turbine or paddlewheel flowmeters, vortex-shedding flowmeters, magnetic flowmeters, differential-pressure flowmeters, and ultrasonic Doppler flowmeters, as discussed in Sections 5.1.1 through 5.1.6. Data on all of the flowmeters discussed are presented in Table 5-1.

5.1.1 Rotameters

Rotameters operate on the principle of variable area where a flowing liquid changes the position of a float, piston, or vane to open a larger area for the passage of fluid (see Figure 5-1). The travel of this moveable indicator is a function of both the flowrate and the viscosity of the fluid. A flowrate measurement is based on a calibrated volumetric scale for

TABLE 5-1. Features of COTS Flowrate Sensors

Manufacturer/ Model	Flowmetrics Series 2000	OMEGA FLR8313-I	MC Meters and Controls 2" Cast Iron Turbine	Fluidyne Model 1100	EMCO Vortex PhD™
Technology	Rotameter—no electronics	Rotameter with electronic output	Turbine	Vortex shedding	Vortex shedding
Range/ Sensitivity	Maximum of 180 gallons per minute (gpm)	10 to 130 gpm	4 to 350 gpm for 2" pipe	2.5 to 125 gpm for 2" pipe	13.8 to 295 gpm for 2" pipe
Accuracy	±2% full scale	2 to 5% full scale	±1.5% full scale	±0.5% full scale	±0.7% of flowrate
External Dimensions	24" L x 4.75" W x 4.75" D 26.5 lb	12.2" L x 4.0" W x 5.4" H For 1.25" pipe	10" L x 6.25" W x 8" H 24 lb	Tee, 5.1" L x ~7" H Flange type, 8.47" L x 9" H	NR
Calibration/ Maintenance Requirements	Factory-calibrated	Factory-calibrated	Factory-calibrated	Factory-calibrated; on-site zero and span adjustment	Factory-calibrated; on-site zero and span adjustment
Cost	Base price ~$850	$1,145	$823	$850	$1,996
Comments	No electronics for remote output; visual only; large device	May be prone to fouling; pipe size goes up to only 1.25"; would need additional electronics for totalizer	Includes turbine, output register, and strainer; conventional brass in-line flowmeter	Cost includes local rate/total output; no moving parts; low flow capacity	Cost includes flow and totalizer outputs; no moving parts; corrosion-resistant

that rotameter's range of operation. Typically, rotameters are used for visual measurements, an aspect that would not lend itself well for use in the RTS. However, since the rotameter technology has been integrated with electronics, process flowmeter information systems that are suitable for remote monitoring applications are now available.

Rotameters offer a wide range of monitoring capabilities from <10 gallons per minute (gpm) to several hundred gpm. These monitoring devices cost about $850 for visual meters to more than $1,000 for electronic meters. These flowmeters are for the higher capacity instruments that might be used at a treatment facility.

Literature was reviewed for 12 rotameter suppliers. For each supplier, the rotameter technology was similar with options such as flowrate, mounting features, electrical or visual displays, and cost. For the sake of reporting, similar options are not repetitively included in the database. Representative rotameter features are presented for two suppliers in Table 5-1.

5.1.2 Turbine or Paddlewheel Flowmeters

Turbine or paddlewheel flowmeters operate on the principle of the moving fluid engaging and spinning a vaned rotor that induces an alternating current in a magnetic pickup coil.

TABLE 5-1. (continued)

Manufacturer/ Model	OMEGA Series FMG 2550	OMEGA Series FMG 94	Kobold RCM Direct Flowmeter	GLI International F1A11B2 Tee-Mount Sensor	OMEGA FD-7000
Technology	Magnetic flowmeter, insertion-type	Magnetic flowmeter, wafer-type	Differential-pressure, direct-reading flowmeter	Impeller-driven sensor	Ultrasonic Doppler
Range/ Sensitivity	0.3 to 20 ft/sec for a 2" to 12" pipe	30 to 300 gpm for 2" pipe	Ranges from 15 to 200 gpm for 2" pipe	12 to 380 gpm for 2" pipe	8-digit LCD in gpm
Accuracy	±2% of reading	±1% of reading	±3% full scale for horizontal flow	1% full scale	±2% full scale
External Dimensions	8.6" diameter x 12.9" H 11 lb	NR 27 lb	6.8" H x 4.55" W x 3.3" D 6.1 lb	Sensor ~5" L x 1.5" diameter. Electronics 5.7" H x 5.7" W x 4.3" D <1 lb	17" L x 11" W x 8" H 23 lb
Calibration/ Maintenance Requirements	Factory-calibrated; 100% keypad calibration	Factory-calibrated; automatic zero reference	Factory-calibrated; on-site zero and span adjustment	Factory-calibrated; on-site zero and span adjustment	Factory-calibrated
Cost	$3,285	$2,562	$1,755	$1,515	$6,500
Comments	Relies on fairly stable fluid conductivity; no moving parts; compact unit; high cost	Relies on fairly stable fluid conductivity; no moving parts; good accuracy and range	No moving parts; 10-pipe diameter minimal laminar flow; valving recommended downstream	Cost includes a tee-mounted sensor, flow monitor/totalizer, and mounting hardware	Noninvasive design; suitable for clear, low-viscosity liquids; includes totalizer feature

NR = not reported.

The pulsed voltage generated by the spinning shaft provides a very accurate measurement of flowrate. These flowmeters offer the following advantages:

- They have a very wide flow range.
- They provide very accurate measurements (for their cost).
- They are constructed of durable materials, allowing for both corrosive and noncorrosive applications.
- They are simple in design and easily maintained.
- They are compatible with a variety of electronic readout devices and controllers.

Turbine flowmeters differ from the paddlewheel design because the turbine generally reflects flow over the cross-sectional area of the pipe (Figure 5-2), whereas the paddlewheel

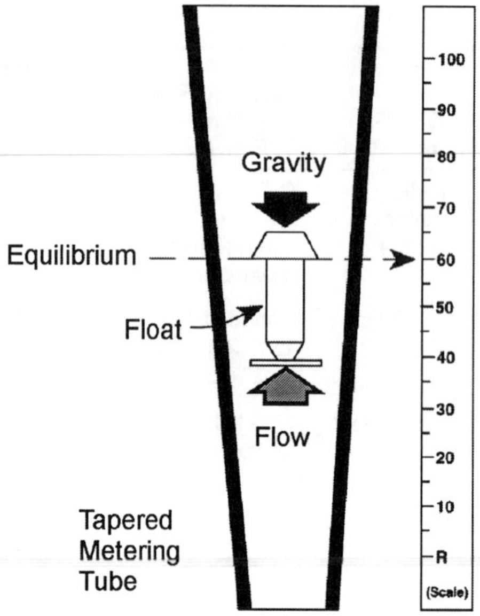

FIGURE 5-1. Operating principle of a rotameter or variable-area flowmeter. (Copyright © *Plant Engineering Magazine*. Adopted from OMEGA Engineering, Inc., *The Flow and Level Handbook*, 1995. Reproduced courtesy of OMEGA Engineering, Inc.)

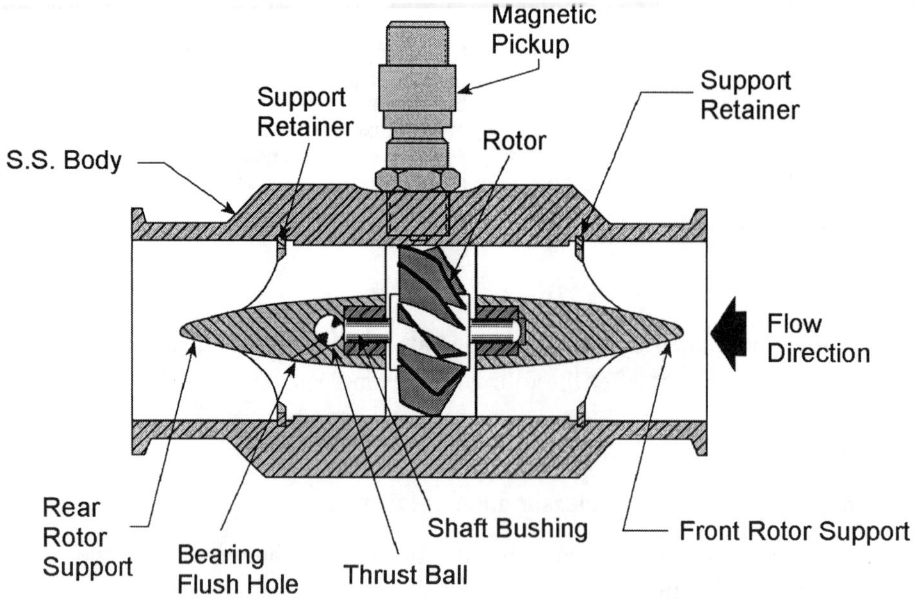

FIGURE 5-2. Schematic of a typical turbine flowmeter. The rotating blades generate a frequency signal proportional to the liquid flowrate, which is sensed by the magnetic pickup and transferred to a readout indicator. (Copyright © *Plant Engineering Magazine*. Adopted from OMEGA Engineering, Inc., *The Flow and Level Handbook*, 1995. Reproduced courtesy of OMEGA Engineering, Inc.)

is affected by only a portion of the fluid moving through a pipe (Figure 5-3). Both devices produce a pressure drop, but of the two, the turbine design has the greater pressure drop, which must be considered when selecting a unit.

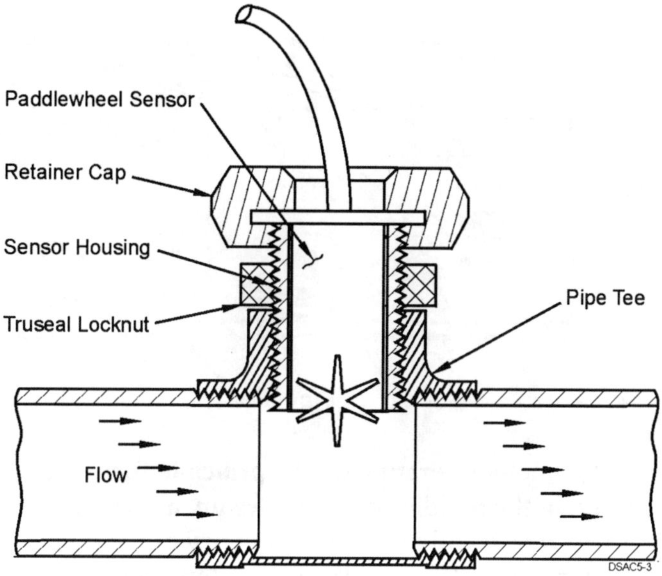

FIGURE 5-3. Typical paddlewheel flowmeter. (Copyright © OMEGA Engineering, Inc., *The Flow and Level Handbook,* 1995. All rights reserved. Reproduced with permission.)

Costs for these mechanical flowmeters can be higher than for rotameters but lower than for the nonmechanical devices that are discussed below. Paddlewheel costs range from about \$200 for the basic unit to about \$500 when electronic packages providing signal conditioning and outputs are included. The turbine design is generally more expensive, because the body of the turbine is built to the diameter of the pipe in which the flow is being measured. These extra material and fabrication requirements raise the cost of turbine meters for 2-inch pipe to about \$1,000 for stainless steel construction, plus about \$800 for output electronics.

The reported accuracy for both of these types of mechanical meters ranges from 1.0 to 5% full scale (FS). This level of accuracy is 2 to 10 times less than that reported for the more expensive flowmeters. If the flowrate is within 1 to 5 gpm and the total flow is less than 100 gpm, this type of flowmeter becomes more cost-effective.

The results of the search for suppliers of turbine and paddlewheel flowmeters identified nine sources. Suppliers offer similar types of turbine and paddlewheel models reflecting various sizes, costs, and accuracy options to perspective buyers.

5.1.3 Vortex-Shedding Flowmeters

Vortex-shedding flowmeters rely on the principle that vortices generated in the meter's body as the fluid passes through will put measurable stress on an internal shedder bar (see

Figure 5-4). These stresses produce measurable pulses that are proportional to the volume of liquid. The shedder bar transmits the stress pulses to an encapsulated piezoelectric sensor that amplifies the signal, producing either a milliamp or a voltage output.

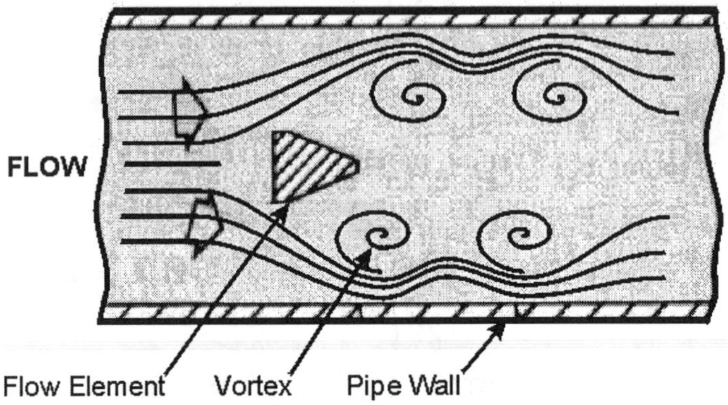

FIGURE 5-4. Vortex flowmeter operating on the principle that when a nonstreamlined object is placed in the middle of a flow stream, a series of vortices are shed alternately downstream of the object. The frequency of the vortex shedding is directly proportional to the velocity of the liquid flowing in the pipeline. (Copyright © *Plant Engineering Magazine*. Adopted from OMEGA Engineering, Inc., *The Flow and Level Handbook,* 1995. Reproduced courtesy of OMEGA Engineering, Inc.)

Flowmeters of this type have no moving parts nor sensor contact with the liquid. Reportedly, this configuration provides accurate, reliable, and low-maintenance flow measurements. The measurement accuracy for liquids ranges from 0.8% to 3% FS for three brands of meters. Flowrates for 2-inch pipe range from about 2.5 to 300 gpm. Flowmeter dimensions are relatively compact, with lengths ranging from 3.2 to 8.4 inches, with a weight of 2.7 lb for plastic-bodied devices and 6.6 lb for metallic construction. Two suppliers of vortex-shedding flowmeters were identified during the search. Prices of the basic units range from $850 to $1,996, with electronic packages costing an additional $775.

5.1.4 Magnetic Flowmeters

Magnetic flowmeters operate on Faraday's law, which states that a voltage induced by any conductor moving at a right angle through a magnetic field is proportional to the velocity of the conductor. For magnetic flowmeters to work, the liquid being measured must be conductive, as this would typically be the case for water at a treatment facility. Two types of magnetic flowmeters were found to be generally in use. They are the insertion type, where a magnetic field radiates outward from a probe (see Figure 5-5), and the wafer style, where a magnetic field is constructed across the entire cross section of the flow tube (see Figure 5-6).

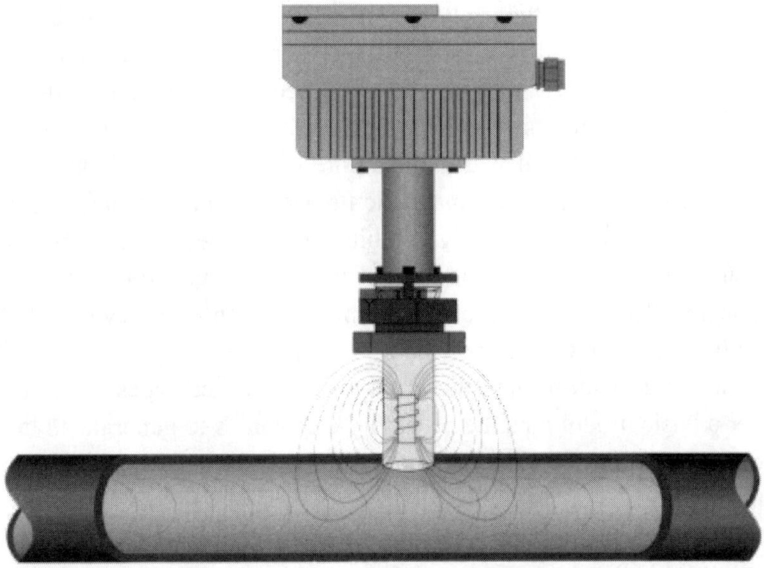

FIGURE 5-5. Insertion-type magnetic flowmeter operating principle. (Copyright © OMEGA Engineering, Inc., *The Flow and Level Handbook*, 1995. All rights reserved. Reproduced with permission.)

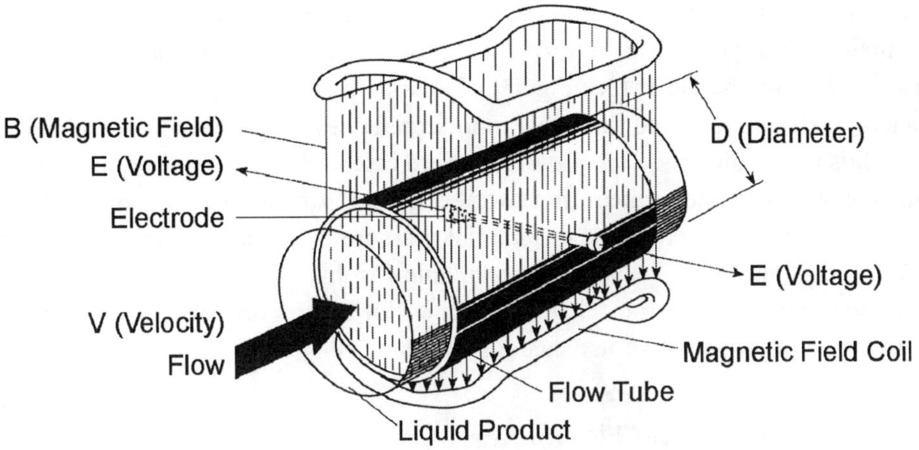

$E \propto VBD$ E = Induced Voltage
V = Average Liquid Velocity
B = Magnetic Field
D = Distance Between Electrodes (Pipe I.D.)

FIGURE 5-6. Wafer-style in-line magnetic flowmeter operating principle. (Copyright © OMEGA Engineering, Inc., *The Flow and Level Handbook*, 1995. All rights reserved. Reproduced with permission.)

Because flowmeters of this type rely on the conductivity of the fluid being measured, several factors need to be considered. First, if the presence of ions varies over time, the conductivity, and therefore the performance, will vary. Second, electrical conductivity is also a function of temperature, so the precise measurement of temperature is needed even though the meters normally are equipped with temperature compensation. Finally, if the conductance decreases due to water treatment, magnetic flowmeters may become less effective.

The wafer-style flowmeter reportedly exhibits a high degree of accuracy, 1.0% FS, and because there are no moving parts nor contact with the liquid, durability and low maintenance are expected. The insertion type of flowmeter has an accuracy of about 2% FS, and one size accommodates pipe diameters from 2 to 12 inches.

These flowmeters cost more than the previously described types. The insertion design costs $3,285 for a basic model plus the necessary electronics to generate a rate display. The wafer type costs $3,016 for 2-inch pipe and would require a $610 electronics package to complete its installation.

Continuous power demands are high for wafer-style magnetic flowmeters, 115 volts alternating current (VAC), whereas the insertion style can operate with continuous 24 volts direct current (VDC). The wafer style permits very-low-flow measurements for 2-inch pipe (0.21 gpm) while allowing maximum full-scale measurements of up to 300 gpm.

5.1.5 Differential-Pressure Flowmeters

The next type of flowmeters reviewed operates on differential pressure. These compact units contain a calibrated, built-in nozzle that produces a measurable pressure drop as the fluid passes through the meter (see Figure 5-7). No moving parts are associated with these flowmeters, and as long as the fluid being monitored does not contain highly abrasive suspended solids that could erode the nozzle, they are essentially maintenance free. For 2-inch pipe, models are available to measure flowrates from 15 to 200 gpm. Compared to turbine or vortex-shedding flowmeters, the differential-pressure flowmeter is lighter in weight and size. Both gauge and digital outputs are available for this flowmeter. The cost of $1,755 makes it more expensive than mechanical turbine flowmeters of the same size, but similar in cost to the nonmechanical vortex-shedding types.

5.1.6 Ultrasonic Doppler Flowmeters

Ultrasonic Doppler flowmeters operate on the principle of measuring the frequency shift (Doppler effect) for an ultrasonic signal reflected by suspended particles or gas bubbles in a fluid stream (see Figure 5-8). This frequency shift occurs when sound waves are reflected by the particles or bubbles in a flowing liquid. The measured frequency shift is proportional to the flow of the liquid. For flowrate measurements to be made, most ultrasonic flowmeters require that at least 25 ppm of 30-micron or larger suspended particles or bubbles exist in the fluid. These are necessary to provide enough "reflectors" so that the ultrasonic signal can be bounced back to the receiving element in the flowmeter. An advantage

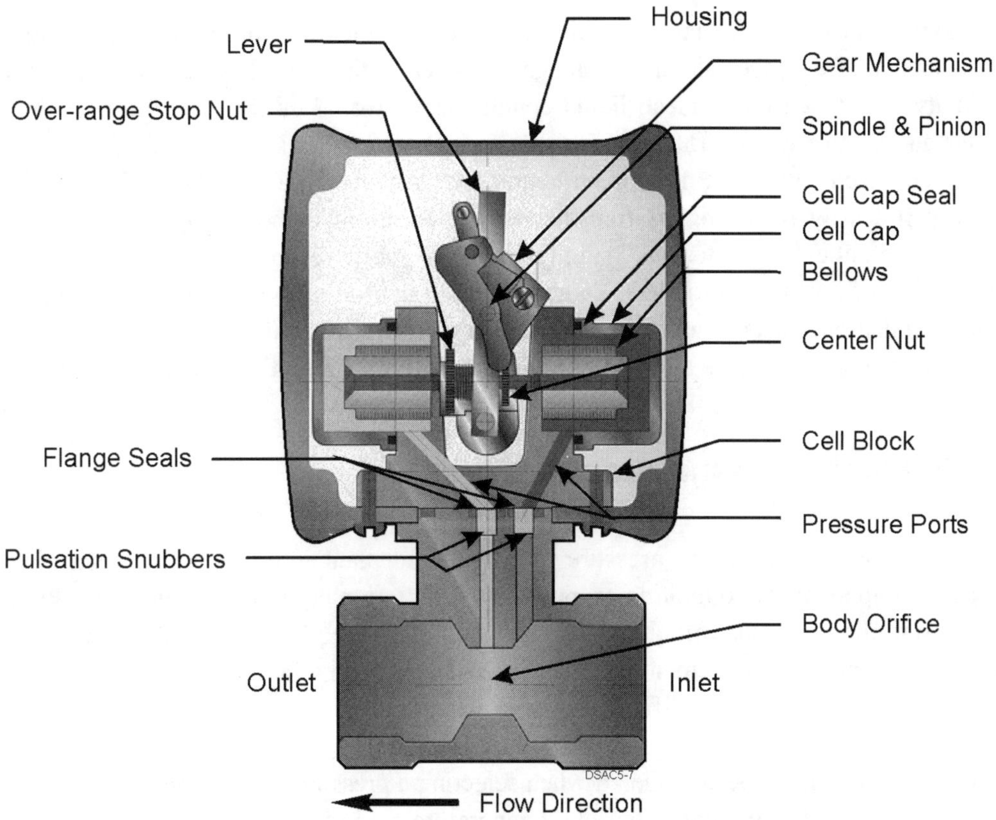

FIGURE 5-7. Typical assembly of a differential-pressure flowmeter. (Adopted from RCM Industries, Inc., "Series 7000 Flo-Gage." Used with permission.)

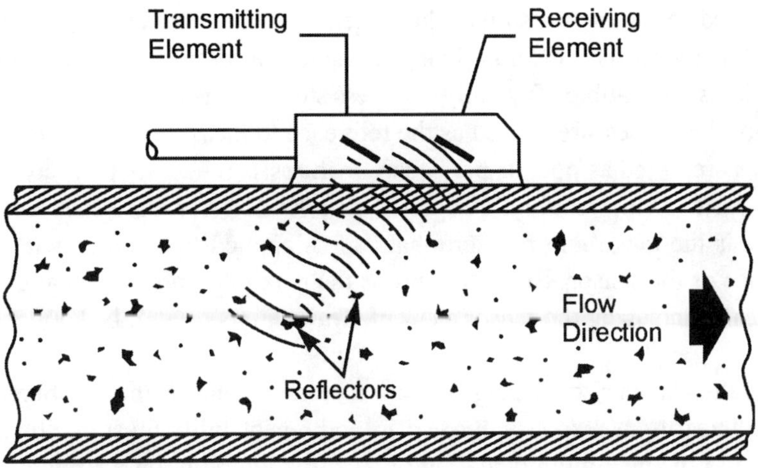

FIGURE 5-8. Typical ultrasonic Doppler flowmeter. (Copyright © OMEGA Engineering, Inc., *The Flow and Level Handbook*, 1995. All rights reserved. Reproduced with permission.)

of this type of flowmeter is that it is noninvasive, because the transmitter/receiver is mounted on the outside of the pipe. The disadvantage, however, is that few flowmeters exist with the sensitivity to be used under clean liquid conditions, as would be the case for groundwater influent and potable water effluent from a groundwater-based drinking water facility.

A specialized ultrasonic flowmeter reportedly can measure the flow of clean, low-viscosity fluids by mounting its transducers at a location in the piping where nonsymmetrical hydraulic turbulence will supply frequency shift response. The microprocessor provides both flowrate and totalizer data. A 4- to 20-mA output would be compatible with a data acquisition package and remote telemetry. The cost of this flowmeter is $6,500, which makes it an expensive alternative to mechanical or differential-pressure options.

5.2 Pressure Measurement Sensors

The measurement of water pressure in a treatment plant is important to determine if pumping components are operating properly and if filtration equipment requires backflushing due to fouling. Pressure monitoring for a typical package plant is expected to consist of either a single-point measurement, such as downstream from a pump or the pressure in the distribution storage tanks, or differential-pressure measurements across a filter or a treatment bed.

Several factors must be considered when selecting a pressure sensor, including the pressure range expected to be monitored, the temperature, and the medium being measured. It is generally acceptable to select a sensor that can withstand 125% of the maximum expected pressure range for the system. This, in conjunction with the rated burst pressure, should permit selecting a sensor that will withstand normal operating conditions and the spikes of pressure typically associated with the startup of pumps. Further, the expected operating temperatures need to be known so that the selected sensor provides temperature compensation across this range. The medium being measured needs to be known so that the diaphragm material is compatible. Typically, stainless steel (SS) is used for a liquid medium.

When atmospheric pressure is used as the reference to measure gauge pressure, the pressure generally is reported as pounds per square inch (psi). When the pressure measurement is relative to vacuum, the reading is reported as pounds per square inch absolute (psia). Proof pressure is the maximum pressure that can be applied to a sensor without changing the performance of the sensor. Burst pressure is the maximum pressure that can be applied without physically damaging the sensor element. These limits need to be known when selecting a sensor.

To identify the accuracy of the signal from a pressure sensor, the combined error associated with its linearity of response, hysteresis, and repeatability must be obtained. Linearity is measured as the maximum deviation of a calibration point on a straight line during a calibration cycle. A pressure sensor's hysteresis is the maximum output difference for a specific pressure when that pressure is approached when first increasing and then decreasing pressure. A sensor's repeatability is the consistency of response when the same pressure is applied under constant conditions and in the same direction.

During the search for pressure sensors, 11 manufacturer/suppliers provided literature. In many cases, similar devices are provided by most of the suppliers. In general, the differences between the pressure sensors reported are the construction features that provide an accurate unit with a wide operating range and acceptable temperature compensation characteristics.

Table 5-2 presents 10 common pressure sensors reviewed, including both single-point measurement transducers and differential-pressure sensors. OMEGA sensors are heavily represented because this company offers a wide range of sensors manufactured by many suppliers that provide options in cost and performance for the variety of conditions expected to be encountered in a water treatment facility. The costs of pressure transducers vary from about \$150 to \$1000. The difference in cost generally is associated with the

TABLE 5-2. Features of COTS Pressure Sensors

Manufacturer/ Model	Ellison Sensors PR9500	MODUS Series W	NOSHOK Series 100	OMEGA PX120	OMEGA PX300
Technology	Silicon-on-sapphire sensor with wireless transmitter	Differential-pressure transducer	Thin-film sensor	Thin-film sensor	Stainless steel diaphragm; silicone oil-filled semiconductor
Range/ Sensitivity	0-500 mbar or 0-1500 bar	0-200 psid static; 0-300 psi	0-100 psig	0-500 psig	0-500 psig
Accuracy	Resolution, 12-bit ADC; linearity 0.3% full scale	±0.5% of DPR	±0.5% full scale	±1% full scale	±0.25% full scale
External Dimensions	Transmitter 156 mm W x 60 mm H Receiver 75 mm D x 120 mm W x 200 mm H	3.56" H x 4.54" W x 2.18" D 1.2 lb	4.5" H x 1.07" diameter	1.875" L x 1.0" diameter 3.0 oz	3.375" L x 1.0" diameter 4.6 oz
Calibration/ Maintenance Requirements	Factory calibration; zero and span adjustments	Factory calibration; field adjustments to zero and span	Factory calibration; zero and span adjustment per calibration data	Factory calibration; field adjustments to zero and span	Factory calibration; field adjustments to zero and span
Cost	Transmitter \$975 Receiver \$750	\$395	\$226	\$158	\$180
Comments	Nonwired application; built-in low-power UHF radio transmitter; no license required	No moving parts; pressure across filters and system; takes place of two pressure transducers	Industrial-grade; singular point measurement; reportedly very stable and long-lived	Industrial-grade; corrosion-resistant	Industrial-grade, general-purpose pressure sensor

TABLE 5-2. (continued)

Manufacturer/ Model	OMEGA PX602-612	OMEGA PX425	OMEGA PX726	Setra 228-1/C228-1	Setra C230
Technology	Thin-film polysilicon pressure sensor	Molecular-bonded silicon strain gauge to sensing steel beam	Stainless steel diaphragm; silicone oil-filled semiconductor	Capacitance differential-pressure sensor	Capacitance differential-pressure sensor
Range/ Sensitivity	0-500 psig	0-15 to 0-6,000 psig	0-10 to 0-5,000 psig	Differential-pressure range 0-100 psid; 1000 psig working range	Differential-pressure range 0-100 psid; 250 psig maximum pressure
Accuracy	±0.4% full scale	±0.19% span	±0.15% of calibrated span	±0.25% full scale	±0.25% full scale
External Dimensions	3.78" L x 1.058" square cap 2.5 oz without cable	2.5" L x 1.17" diameter 6 oz	5.82" H x 4.4" D 1 lb 9 oz	3.05" D x 2.7" diameter ~2 lb	3" H x 2.84" D x 2.15" W 14.4 oz
Calibration/ Maintenance Requirements	Factory calibration with span data	Factory calibration; field adjustments to zero and span	Factory calibration; field adjustments to zero and span	Factory-zeroed after application of 300 psi to high side	Factory-zeroed after application of 300 psi to high side; must be rezeroed after use
Cost	$198 + $10 snubber	$495 + $10 snubber	$710 + $10 snubber	$870	$250
Comments	General-purpose pressure transducer; excellent long-term stability	Excellent long-term stability; shock- and vibration-resistant	Industrial mounting in NEMA box; conduit fittings	Suitable for determining pressure drop across filters under high working pressures	Suitable for determining pressure drop across filters under moderate working pressures

ADC = analog to digital conversion.
D = depth.
DPR = differential pressure range.
H = height.
L = length.
mbar = millibar.
NEMA = National Electrical and Mechanical Association.
psi = pounds per square inch.
psid = pounds per square inch, differential.
psig = pounds per square inch, gage.
UHF = ultrahigh frequency.
W = width.

sophistication of the electronics for temperature compensation, signal conditioning, and wireless data transmission.

5.3 Finished Water and Wastewater Production Flowmeters/Totalizers

Tracking the volume of water produced, as either finished water or wastewater, is accomplished by totalizers in conjunction with flowmeters. Totalizers can be either mechanical or electronic devices which, in the case of a water treatment facility, maintain a running total of the volume passing a monitoring point in the water treatment systems. If the device is located at the backside of the pump that is delivering the water to the tank for distribution, a water production value is generated. If it is located on the effluent from either a backflushing operation or an NF/RO reject stream, wastewater volume is monitored.

Mechanical totalizers are simple gear-driven dials that advance as a flowmeter responds to the volumetric movement of water past a point. An example of this type of totalizer is the conventional water meter found in households. An electronic totalizer is a microprocessor that "counts" signals produced by a flowmeter, generating a digital readout of the volume that passes a monitoring point (Figure 5-9).

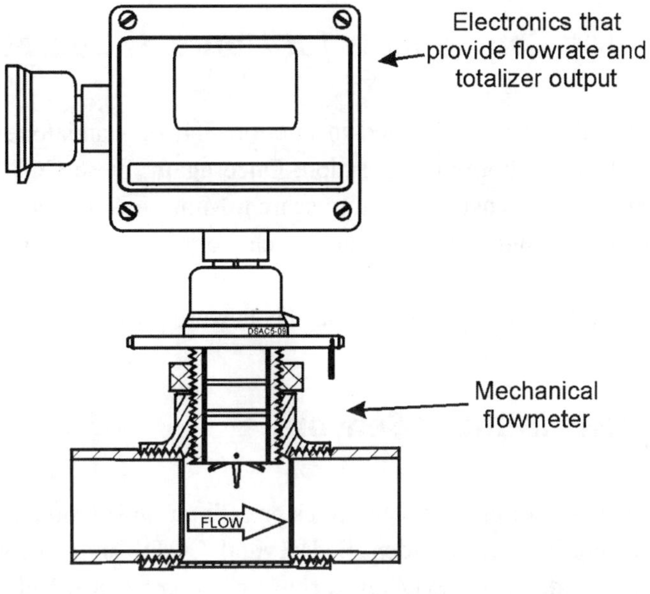

FIGURE 5-9. A typical electronic flowmeter/totalizer that can be interfaced with flow-monitoring devices. (Adopted from GLI International, Inc., Data Sheet 680F, 1997. Used with permission.)

The literature search indicated that most flowmeters have the option of being fitted with either a mechanical or a digital totalizer. The electronic totalizers are of primary interest for this task because they provide output that can be incorporated into an RTS and accessed remotely. Totalizer design produces either a millivolt, volt, or milliamp signal that would be compatible with DAQ systems.

Typical standalone electronic components that can be interfaced to a flow-monitoring device cost about $1,000. These digital counters have the capability of being used not only for flow but also for electrical usage. For ease of application and replacement, flowmeter/totalizer packages provide an attractive component for an RTS. Totalizer electronics data are included in Table 5-1 for flowmeters.

5.4 System Status Electronics

A literature search for specific electronic components to monitor system status was not performed. A search with keywords including "electronics" and "system status" would have resulted in an oversized, counterproductive printout. Numerous electronic suppliers have simple relays for monitoring continuity and functionality of electrical devices at a package water treatment plant. Components generally are inexpensive, have rapid response characteristics, can be networked across a package plant, and are easily integrated into a remote-monitoring system. No tables or data sheets were prepared for these COTS electronic items.

5.5 Monitoring Components for Hours of Operation

Programmable relays, as mentioned in Section 5.4, or standalone timers/counters are readily available to track hours of operation. Once again, these COTS electronic components did not require an intensive literature search. Multiple suppliers of these components were identified in electronics and instrument catalogs. Timers/counters are inexpensive, ranging from $40 to $200, and include features that can provide remote tracking and on-site visual digital readouts. No tables or data sheets were provided for these COTS items.

5.6 Level Measurement Sensors

Monitoring the levels of chemical additives or calibration solutions is expected to be part of the RTS. The literature search identified several COTS options for monitoring levels, including cable-suspended float switches, mechanical floats that travel up and down a shaft containing reed switches, rigid or flexible electrodes, integrated silicon strain gauge bridge pressure transducers, ultrasonic and acoustic level transmitters, nonintrusive capacitive switches, and radiofrequency capacitive level transmitters, as described in Sections 5.6.1 through 5.6.7. Six manufacturers/suppliers were identified. There are undoubtedly other manufacturers/suppliers of level sensors, but for the six that provided literature, similar technologies of level sensing were identified and suitable components to perform this monitoring are available. Table 5-3 presents 12 level-monitoring options identified during the search.

5.6.1 Mechanical Cable/Float Devices

The first level-monitoring option relies on the density of the fluid to displace a float which, in turn, is electronically integrated into microswitches. The position of the float adjusts the angle of the switches, which can indicate a "high" or a "low" level condition or can trigger a "pump on" or a "pump off" condition to refill a tank (Figure 5-10). These types of level sensors generally are not affected by materials that generate a conductive film on

TABLE 5-3. Features of COTS Level Sensors

Manufacturer/ Model	Kari Model M2HL	Kobold NM-298	OMEGA LVR-30	GLI International 4000L	GLI International 4300L	In-Situ, Inc. PXD-260
Technology	Cable-suspended float switches	Mechanical float with reed contacts	Magnetic float/reed switches, voltage divider network	Electrode conductance, rigid	Electrode conductance, flexible cable	Integrated silicon strain gauge bridge, pressure transducer
Range/ Sensitivity	Standard cable length of 16'	0"-36", additional 4" section available	0'-6'	0'-19.5' in 1-inch increments	0'-12', in 1-inch increments	0'-578' depending on unit selected
Accuracy	10" minimum switching differential, 40" maximum	±0.6"	±0.4% full scale or ±0.5", whichever is greater	0.5% full scale	0.50% full scale	±0.15% of range-linear coefficients
External Dimensions	Float: 3" diameter x 4.85" L	Float 1.7" diameter x 1.8" long; shaft up to 6.6' L	2" NPTM fitting	Electrode diameter, 3/8" with lengths up to 19.5"	Cable diameter 3/8" with lengths up to 12'	Probe head is 1" OD x 10.9" length
Calibration/ Maintenance Requirements	Factory-adjusted contact switch; field adjustments of cable length for alarm settings	Factory-calibrated control instruments	Factory-calibrated	Factory-calibrated	Factory-calibrated	Factory-calibrated
Cost	$127 for float/cable assemblies	Sensor $619; shaft $304; signal conditioner $222	Sensor $458, DPF50 process meter $160	Sensor probe ~$200; level transmitter $590	Sensor cable ~$300; level transmitter $590	~$1,100
Comments	Fluid specific gravity of ≥0.95 required	Requires fluid specific gravity of ≥0.8	Chemically resistant materials; liquid must have specific gravity of ≥0.75	Not recommended for materials that form a conductive film	Not recommended for materials that form a conductive film	Stainless steel probe; generally used to monitor groundwater fluctuations but usable for tanks

TABLE 5-3. (continued)

Manufacturer/ Model	Kobold NEO-5000	OMEGA LVU-300	Kobold NEA-2	INOTEK Model 8206	OMEGA LVC-101	OMEGA LV-5200
Technology	Ultrasonic level transmitter	Continuous, ultrasonic sensor	Acoustic level transmitter	Nonintrusive capacitance switch	Intrusive radiofrequency capacitance sensor	Radiofrequency capacitive level measurements
Range/ Sensitivity	0.5' to 24.5'	Maximum of 24.5' from sensor face	0.5' to 4'	Measures upper and lower limits in a tank	Based on cable length and sensor position. Standard, 0'-8'	Maximum length of 12'
Accuracy	±0.25% full scale	±0.25% full scale	±1% span (in air)	±1.0 mm in water	±1 mm in water	±0.5% full scale
External Dimensions	3" L x 2" diameter	Sensor housing 2 x 2.2"; electrical housing 2.8 x 4.5"	Sensor head 2.8" diameter x 4.5" L; waveguide tube 48"	2.5 x 1.7 x 1.5"	3" L x 0.75" diameter	Sensor gland 3/4" NPT, transmitter 6.9 x 4.3"
Calibration/ Maintenance Requirements	Calibrated on site with span setting	Factory-calibrated	Fixed offset, automatic temperature compensation	Not required because sensor acts as on/off switch	Factory calibration; no field efforts required	Transmitter requires span range and zero range adjustments
Cost	$850	Sensor $895, signal controller $425	$449	Sensor $195	LVC-101 sensor $185, LVM-13 6' mounting track $120	Sensor $180 Transmitter $655 Accessories ~$600
Comments	Noncontact measurement; automatic temperature compensation	Noncontact measurement; automatic temperature compensation; controller permits alarms, multiple setpoints	Not affected by vapors, turbulence in the tank; low interference problems because wave travels in protective waveguide	Located on outside of nonmetallic tanks; indicates if tank is full or empty	Sensor is a switch that indicates if it is submerged or exposed to air; impervious to changes in temperature, pressure, light, particulates, color or density of liquid	Conductive film-resistant coating on the element; transmitter can be located remotely

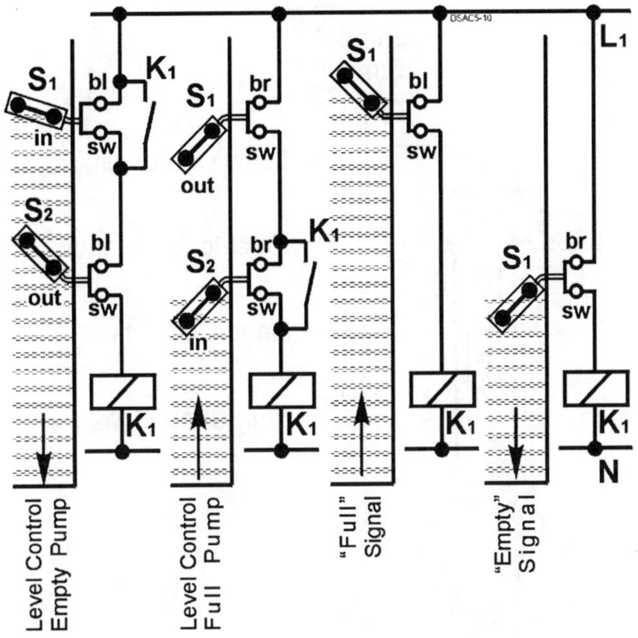

FIGURE 5-10. Float switch configuration for level monitoring and control. (Adopted from Kobold Instruments, Inc., *Flow, Pressure, Level, Temperature: Measuring, Controlling, Regulating,* ©1997. Used with permission.)

the float, and they require little or no maintenance. A single float assembly costs about $125 and can be used over a wide range of tank depths by increasing the cable length.

These monitors require at least 10 inches of fluid to activate the switch and a maximum switching differential of 40 inches for high and low conditions. Also, the specific gravity of the fluid must be greater than 0.95 for the floats to operate properly. Depending on the float design selected, either multiple float/switches or a single multilevel switch can be used for level sensing.

5.6.2 Magnetic Floats/Mechanical Reed Switch Systems

The magnetic float/mechanical reed switch systems operate on a principle similar to that of cable/float devices except that, for the former, the float rides up and down on a rigid shaft as the fluid level fluctuates. The float's location on the rigid shaft is indicated electronically by the magnetic opening and closing of a series of reed switches in the shaft (Figure 5-11). These level sensors are chemically resistant and, as with the cable/float devices in Section 5.6.1, they rely on the specific gravity of the fluid being measured to displace the float. The accuracy of magnetic float/mechanical reed switch sensors is reportedly high, with level changes of ±15 mm possible. The cost of a typical base unit is about $600 for the sensor, $38 per 4-inch increment for the rigid shaft beyond the initial 36 inches, and about $200 for the electronics that provide readout and signal output.

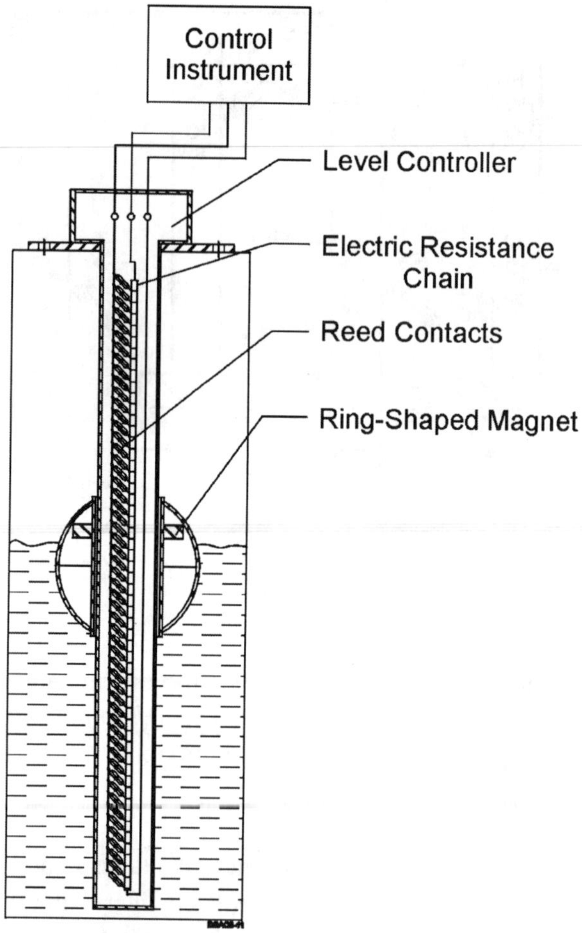

FIGURE 5-11. Magnetic float/mechanical reed switch level sensor. (Adopted from Kobold Instruments, Inc., *Flow, Pressure, Level, Temperature: Measuring, Controlling, Regulating,* ©1997. Used with permission.)

5.6.3 Rigid or Flexible Electrodes

Electrode capacitive sensors can be either rigid probes or flexible cables. Depending on the nature of the fluid measured, either the probe and the tank wall (nonconductive liquid) or the probe and the fluid (conductive liquid) form the two plates of a capacitor. Figure 5-12 shows an example of capacitive measurement in a nonconductive medium. This configuration reportedly can provide very accurate continuous monitoring of changes in fluid level, but these sensors generally are not recommended for materials that may form a conductive film on the probe. Depending on the conductive nature of the fluid and the materials used to produce the sensing probe, it may be possible to employ this design in an RTS. The accuracy of these types of sensors is reported at about 0.5% FS of the device. With electronics, they cost from about $800 to $900 each.

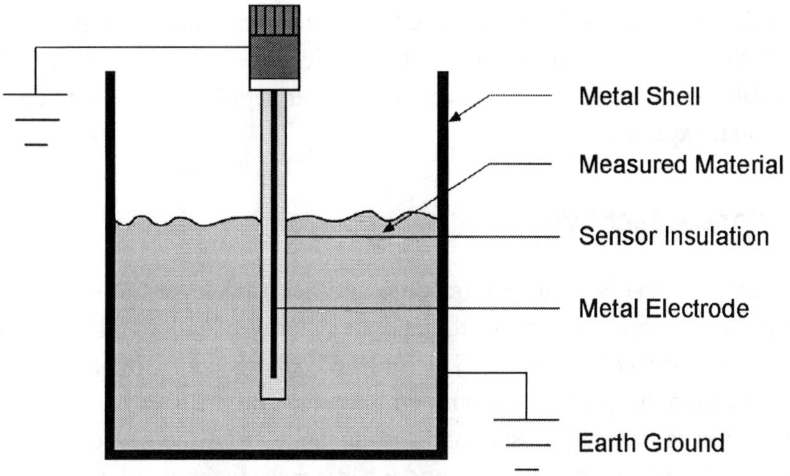

FIGURE 5-12. Example of capacitive measurement in a nonconductive medium. (Copyright GLI International, Inc. Adopted from OMEGA Engineering, Inc., *The Flow and Level Handbook,* 1995. Reproduced courtesy of OMEGA Engineering, Inc.)

5.6.4 Integrated Strain Gauge Pressure Transducers

Integrated strain gauge pressure transducers rely on the pressure exerted by the fluid head to determine the level. The pressure is continually compared to atmospheric pressure via a pressure equalization tube that is always open to the atmosphere. The devices typically are used to monitor fluctuations in groundwater (see Figure 5-13), but the technology also

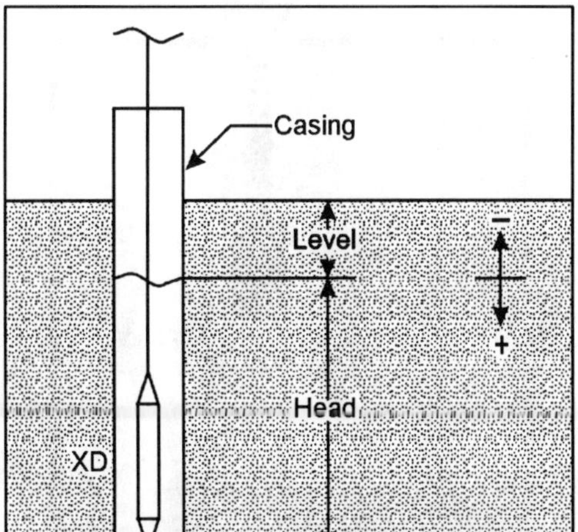

FIGURE 5-13. Example of an integrated strain gauge pressure transducer for level sensing. (Adopted from In-Situ, Inc., Pressure Transducer Model PXD-260 Operator's Manual, 1994. Used with permission.)

is applicable for tank level sensing. The transducers generate a signal that provides measurements accurate to ±0.15% of range-linear coefficients. They cost about $1,100 for the transducer, cable, and electrical connections to a power supply. The readout electronics would be an additional expense.

5.6.5 Noncontact Sensors

Noncontact sensors include ultrasonic and acoustic level transmitters. Ultrasonic level transmitters use a microprocessor to generate a series of pulses that are directed toward the surface of the fluid (Figure 5-14). The returned echo is compared to a signal gate produced at the same time the pulse is transmitted and the time differential is determined. This temporal function is then processed, and the distance from the sensor to the fluid surface is calculated. The advantages of this noncontact sensor technology are that it is not fouled by the fluid being monitored, and it reportedly can provide accurate, long-term service. Measurements can be affected by tank turbulence and vapors.

FIGURE 5-14. Operation of a noncontact ultrasonic level sensor. (Copyright © OMEGA Engineering, Inc., *The Flow and Level Handbook,* 1995. All rights reserved. Reproduced with permission.)

Acoustic level transmitters also rely on wave transmission; however, they utilize a waveguide to maintain wave integrity, thereby solving many of the problems associated with the ultrasonic level-detection technology. Figure 5-15 presents an example of an acoustic level transmitter with a waveguide. Ultrasonic and acoustic sensors identified during the search provide accuracy of ±0.25% and 1.0% FS, respectively, and cost from about $450 to $1,300 for complete units.

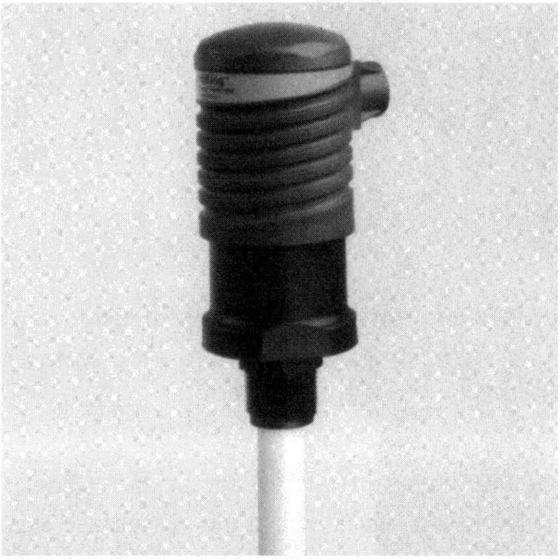

FIGURE 5-15. Acoustic level transmitter with waveguide. (From Kobold Instruments, Inc., *Flow, Pressure, Level, Temperature: Measuring, Controlling, Regulating,* ©1997. Used with permission.)

5.6.6 Nonintrusive Level Sensors

A nonintrusive level sensor is a capacitance switch that can be installed on the outside of nonmetallic tanks. Figure 5-16 shows the exterior (a) and the interior (b) of a nonintrusive radiofrequency (RF) capacitance sensor. This type of sensor can detect the presence of fluid in plastic or fiberglass tanks that have wall thicknesses of up to 1 inch. Such sensors reportedly can provide accuracy levels of ±1 mm in water and are easily retrofitted to existing tanks without additional fittings. An individual sensor would have to be used for each "high" and "low" level measurement. Each of these units costs about $200 and would require additional electronics to be interfaced to an RTS.

5.6.7 Intrusive Radiofrequency Capacitance Sensors

The last type of level sensor identified is an intrusive RF capacitance sensor. This sensor relies on its ability to respond to being located in either air or fluid. A single sensor needs

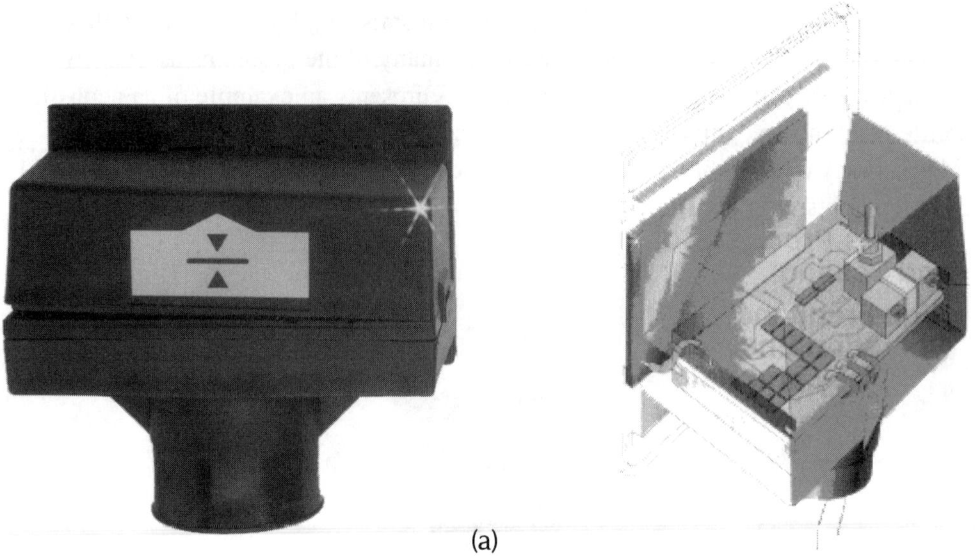

(a) (b)

FIGURE 5-16. Exterior (a) and interior (b) of a nonintrusive capacitance sensor. (Copyright © OMEGA Engineering, Inc., *The Flow and Level Handbook,* 1995. All rights reserved. Reproduced with permission.)

to be located at each of the depths of interest in the tank. These sensors are mounted sequentially on a plastic track and suspended in the vessel (see Figure 5-17). Accuracy is reportedly ±1 mm in water. Each sensor costs about $185, with a 6-foot-long mounting track costing $120. Reportedly, these sensors are impervious to changes in temperature, particles, fluid density, or color.

Level sensor options for the RTS generally reflect what was observed for the other process-monitoring sensors. Several suitable technologies are available to perform level measurements, but their costs increase with their sensitivity.

5.7 Remote Telemetry System Pumps and Valves

The RTS may include pumps and valves. Pumps may be used to deliver calibration solutions to a flowthrough cell during validation of sensor/probe performance. Valves may be needed to direct water samples to the same sensor/probe suite, isolate water samples for analysis, and permit water to be directed to waste. The pumps and valves presented in Sections 5.7.1 and 5.7.2, respectively, are not the large-scale versions that typically would be associated with the transfer and delivery of source water from wells to storage tanks.

5.7.1 RTS Pumps

The pumps used to deliver water or calibration solutions to an RTS flowthrough cell should permit variable delivery rates while providing a tolerance to a wide range of viscosities

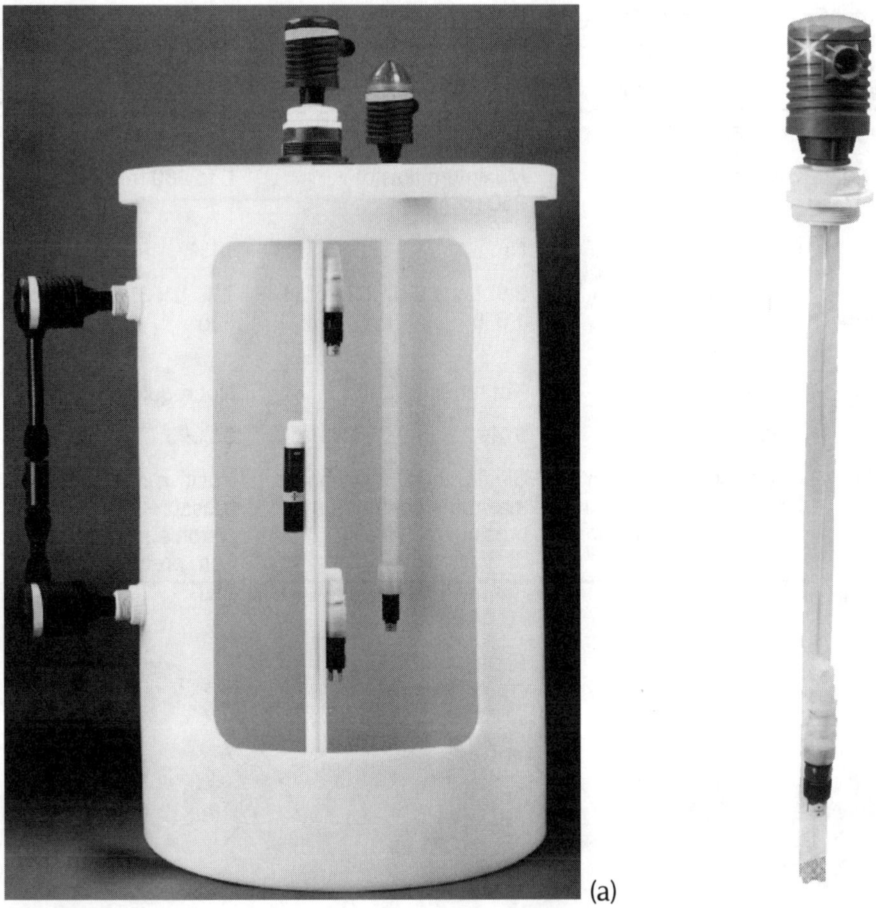

(a) (b)

FIGURE 5-17. (a) Level monitoring and control with intrusive capacitance sensors. (b) Typical capacitance sensor. (Copyright © OMEGA Engineering, Inc., *The Flow and Level Handbook*, 1995. All rights reserved. Reproduced with permission.)

and chemical properties. Three of the available options are peristaltic, diaphragm, and piston-type pumps. Information on these types of pumps is summarized in Table 5-4.

Peristaltic pumps rely on the principle of a rotating roller head to compress tubing filled with fluid, causing the fluid to move (see Figure 5-18). The fluid comes into contact with only the tubing. The peristaltic pump can deliver various output flows depending on the diameter of the tubing and the revolutions per minute (rpm) of the roller assembly. For example, one peristaltic-type pump can deliver a flow as low as 3 mL/min with $^{1}/_{16}$-inch tubing at 14 rpm and up to 360 mL/min with ¼-inch tubing at 108 rpm. This pump costs $198 and may be suitable for delivering calibration solutions from a storage vessel to an RTS flowthrough cell. Peristaltic pumps can be easily operated and maintained.

The second type of pump is the microdiaphragm pump, which can move both gases and liquids. This pump operates by having a diaphragm, in conjunction with a set of check valves, flexed by a cammed armature connected to a motor to form a slight vacuum. The

TABLE 5-4. Features of COTS Remote Telemetry System Pumps

Manufacturer/ Model	VWR MANOSTAT	KNF Neuberger, Inc. Model NF30KN	Fluid Metering, Inc. Model QV-RH1-CKC-W
Technology	Peristaltic pump	Microdiaphragm pump	Valveless piston pump
Flow Range	3 to 360 mL/min	Maximum flow of 250 mL/min	0 to 180 mL/min
Accuracy	NR	NR	±1%
External Dimensions/ Weight	5" L x 6 $^{9}/_{16}$" D x 5 $^{5}/_{16}$" H 5 lb	2.9" L x 1.2" D x 2.25" H 1.3 lb	7⅛" L x 3" D x 3" H 8 lb
Calibration	Not required	Not required	Not required
Cost	$198	$149	$1,095
Comments	Designed for constant flow; 25 psi continuous and 40 psi intermittent delivery pressures	Single-head, micro pump, maximum continuous pressure of 15 psig	Positive-displacement, high-pressure (100 psig) pump; piston susceptible to scoring from precipitated salts

NR = not reported.

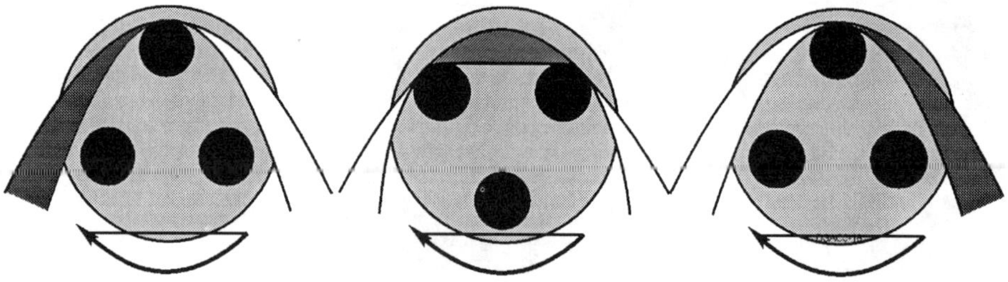

FIGURE 5-18. Basic operation of a peristaltic pump. (Copyright © Watson Marlow Bredel Pumps. Adopted from OMEGA Engineering, Inc., *New Horizons in Fluid Handling for Research and Industry*, 1997. Reproduced courtesy of OMEGA Engineering, Inc.)

vacuum draws gas or fluid into the pump head and forces it out with the next production stroke. One supplier manufactures a microdiaphragm liquid pump that can deliver a flowrate of 250 mL/min with a maximum continuous head pressure of 85 psig. The pump is less than 4 inches long and 3 inches tall, and it weighs only 1.3 lb. The cost of a pump with a polypropylene head and neoprene diaphragm and valves is $149. The compact size, performance characteristics, and cost make this type of pump attractive for incorporation into the RTS to transfer calibration solutions and provide rinse cycles to the flowthrough cell.

The third type of pump is the valveless piston pump, which operates by a cam-driven piston. The piston draws liquid into the pump head on the vacuum stroke and compresses the liquid on the pressurization stroke, which forces the fluid to the delivery point (see Figure 5-19). One valveless piston pump offers a wide range of operation by changing the stroke of the piston to reduce or increase fluid flow. This positive-displacement design can provide higher delivery pressures than the peristaltic pump, but at an increased cost. A pump

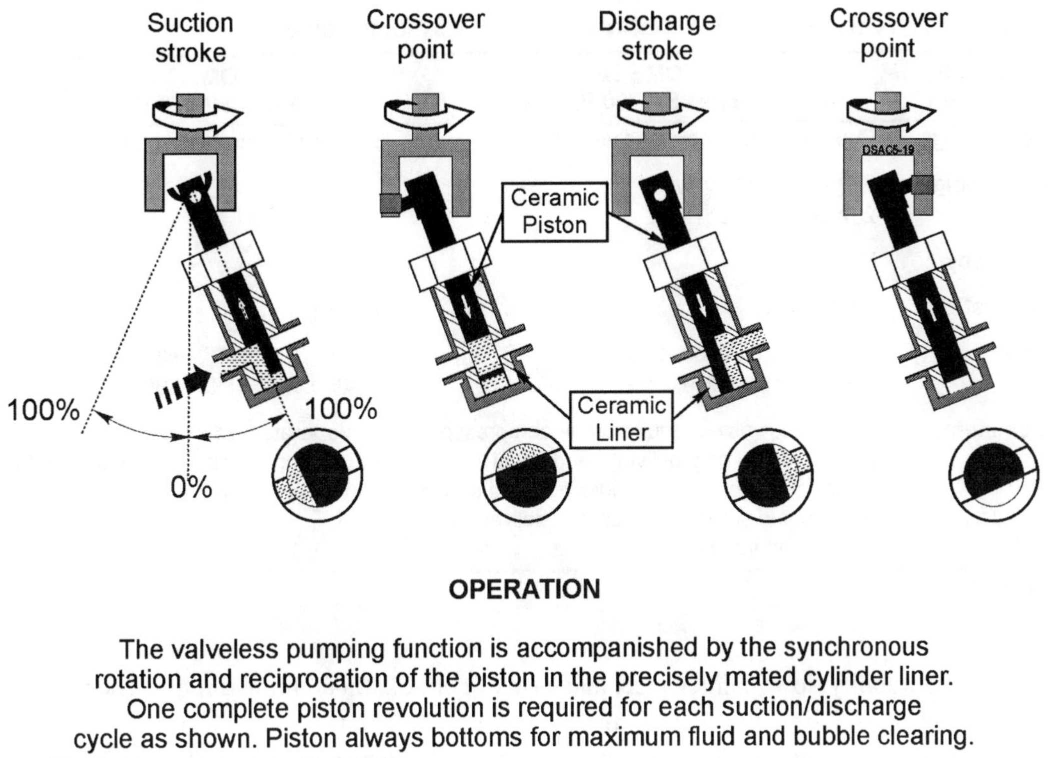

FIGURE 5-19. Operation of a valveless piston metering pump. (Adopted from Fluid Metering, Inc., *Valveless Metering Pumps and Dispensers*, ©1996. Used with permission.)

equipped with a ceramic pump head, an electronic controller, and an isolation and temperature gland to help prevent the precipitation of salts from the fluid costs $1,095.

Because of their simplicity of operation and ease of maintenance, these pumps would be suitable for incorporation into the RTS calibration fluid-delivery subsystem.

5.7.2 RTS Valves

Valves may be needed to direct potable water into the RTS, isolate samples in the flow-through cell, and release water from the cell to a waste containment vessel. Representative valves are shown in Table 5-5. Manually operated valves, such as simple ball valves, may be incorporated into an RTS as backup components, but valves actuated either pneumatically or electrically are the most useful for remote operations.

Pneumatic valves generally rely on a source of air pressure to drive the valve gate from the open to the closed position. The valve may be spring-loaded to maintain a "normally closed" position which, when overcompensated with air pressure, allows the gate to be opened. Once the air pressure orifice is deactivated, the spring returns the valve to a closed position. Pneumatic valves are commonly used in industrial processes, where they have

TABLE 5-5. Features of COTS Remote Telemetry System Valves

Manufacturer/ Model	OMEGA Series SV-700 Valves	OMEGA Series SV-350 Valves
Technology	Pneumatic diaphragm valve	Electrically activated solenoid valve
Flow Range	22 gpm for 0.5" valve 232 gpm for 2.0" valve	NR
Accuracy	NR	NR
Calibration	Not required	Not required
Cost	SV-701, 0.5" orifice, $222 SV-706, 2.0" orifice, $490	SV-351, 0.5" NPT, $98 SV-359, 2.0" NPT, $560
Comments	Pneumatic valves require an air compressor with 80 to 145 psig delivery, additional components for position indication and stroke limiting, and solenoids to control delivery of pressure to the valve	Can operate on 12 or 24 VDC; will require contact closure electronics to actuate the valve

NR = not reported.

exhibited durability, low maintenance, and ease of control. An additional requirement, however, is their need for a compressor/pressure reservoir system to supply air pressure to drive the valve. This adds another complete system to the RTS, which would result in an increased cost beyond the valve itself and would introduce an additional level of maintenance.

One example of a pneumatic valve is the piston-operated diaphragm valve (see Figure 5-20). This valve is designed to be installed in plastic systems having pipe sizes ranging from ½ to 2 inches, with flows ranging from 22 to 232 gallons per minute, respectively. For "normally closed" circuit functions, a minimum pressure of 79.8 psi is needed to drive the valve to an open position, with a maximum pressure rating of 145 psi. The cost of these pneumatic valves ranges from $222 for the ½-inch valve to $490 for the 2-inch valve, with

FIGURE 5-20. Example of a pneumatically driven diaphragm valve. (Copyright © OMEGA Engineering, Inc., *The Flow and Level Handbook*, 1995. All rights reserved. Reproduced with permission.)

additional option costs of \$220 for electrical position indicators and up to \$73 for stroke limiters, to permit flow adjustments. Further costs of approximately \$250 would be incurred for a compressor/pressure tank assembly.

An electrically activated solenoid valve is another option for directing water flow to the RTS. These valves can be purchased in either "normally open" or "normally closed" position. They operate under the same principle as the pneumatic valve, with spring tension overcome by another force to actuate the valve to permit water flow. In this case, an electrical current generates a magnetic force to pull the valve gate open (see Figure 5-21 for a cross section of an electrically activated solenoid valve). Solenoid valves are widely used in process systems, and they range in cost from \$98 for ½-inch brass pipe units to \$560 for 2-inch brass pipe units. These types of valves would be easily controlled by the RTS to allow water to flow into and out of a sensor flowthrough cell.

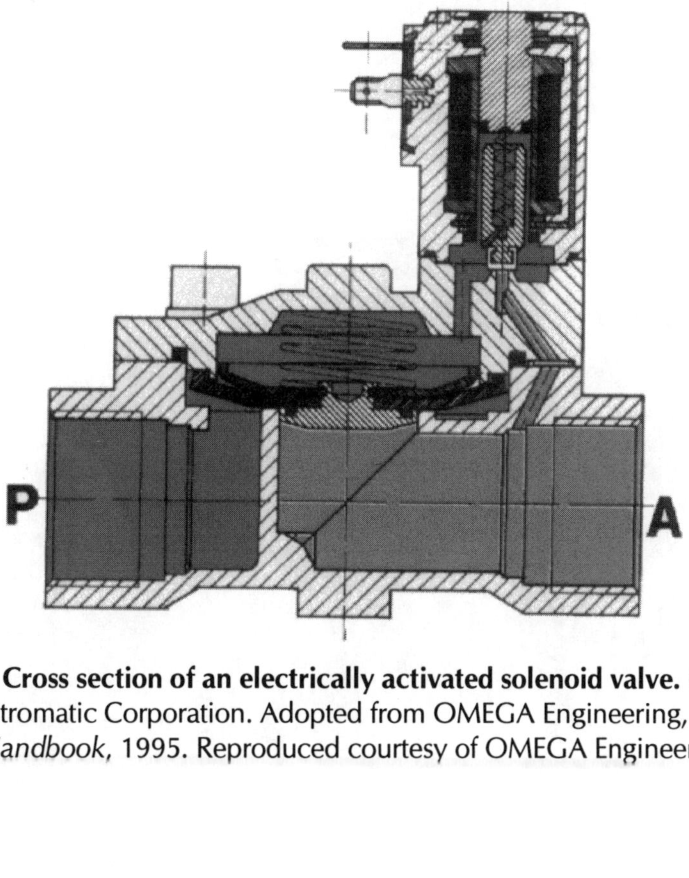

FIGURE 5-21. Cross section of an electrically activated solenoid valve. (Copyright © Burkert Contromatic Corporation. Adopted from OMEGA Engineering, Inc., *The Flow and Level Handbook*, 1995. Reproduced courtesy of OMEGA Engineering, Inc.)

6

COTS Data Acquisition System Options

This chapter introduces data acquisition (DAQ) design issues and presents information on currently available hardware. This chapter is not intended to be an exhaustive treatise on DAQ techniques or related hardware because those subjects are well beyond the scope of this publication. Entire books have been written on DAQ systems, and many vendors sell and support hardware and software used to acquire, store, and display data. Furthermore, hardware and software developments are occurring so rapidly that any attempt to benchmark them would be dated upon publication of this book. Therefore, this chapter is intended as an overview of DAQ systems. Vendors should be consulted for current hardware, software, and recommendations for a specific system configuration.

6.1 Overview of Data Acquisition Systems

DAQ systems are designed to acquire, store, and display data. Currently, these functions are accomplished most often with a digital computer, associated signal conditioning, and software for handling and storing the data.

The main reason for monitoring remote package treatment plants is to electronically track what is occurring at the facility and to permit taking corrective actions when parameters have exceeded specified limits. To do this, the DAQ system, in addition to being able to acquire data, must be able to process the data and provide output signals to the appropriate control mechanism. The control outputs from a DAQ system generally fall into one of two categories, analog or digital. An analog output, like an analog input, can fall within a range of values. Typical ranges for a DAQ system are +10 to −10 VDC or 4 to 20 mA. The second category is digital. Digital inputs and outputs (I/O) are either "on" or "off."

An example of an analog input/output would be a water flowrate monitor and control operation. An analog signal from a sensor measuring flowrate would be acquired by the DAQ system. The processor function inside the DAQ system would compare the measured flowrate with an acceptable flowrate. If the measured flowrate is too low, a command would be generated and an output would be sent from the DAQ system to increase flow by opening the proportional valve that controls flow. The inverse would be possible if a high-flowrate condition is measured. The DAQ system would correct the situation by automatically closing down the proportional valve.

Using the same water flow example to illustrate a digital system, a sensor would determine only if flow exists. No indication of the amount of flow would be available, only "yes, there is flow" or "no, there is no flow." If the processor determines that there should be no flow, it would send a command to close the valve. Therefore, the only commands available to the valve under digital control are "full closed" or "full open."

6.2 Data Acquisition System Design and Component Selection

When designing a DAQ system, the following components that comprise the system should be considered:

- **Sensors.** These devices convert an engineering measurement into an electrical signal that can be used by the DAQ system. Examples are pressure transducers, thermocouples, limit switches, and water quality or process monitors.
- **Signal Conditioning.** The electrical signals generated by analog sensors are sometimes small and susceptible to noise contamination. Signal conditioning is used to amplify or modify the signal into a more robust or usable form. Many sensors have signal conditioning built into them.
- **Analog to Digital Conversion (ADC).** For analog signals to be used by digital computers, they must be converted to digital format. Many vendors build systems and boards to accomplish this conversion.
- **Data Presentation.** Typically, data are presented on monitors or routed to printers for operator access. Software to handle the presentation of data is readily available.
- **Data Storage.** Data are recorded and available for review or trend analysis.
- **Control Algorithm.** A control method is used only when there are digital or analog outputs. Control decisions can be based on a single input, a combination of inputs, or temporal functions such as the time of day or day of the week.
- **Digital to Analog Conversion (DAC).** For analog outputs, the digital information must be converted to an analog signal that can be used by the actuator, often a voltage or current signal.
- **Actuators.** These are the devices that actually affect or control operations. Valves, pumps, motors, and heaters are examples of actuators.

Many vendors sell DAQ systems that are ready to install. Choosing a DAQ system or designing a custom system from COTS components is not a trivial task. The first step in any design is to make a complete list of all the sensor inputs and control outputs that will be required. When compiling these I/O requirements, it is important to remember that the total cost of the system will rise with the addition of each channel. The total channel count should be minimized to the extent possible. Once a list of I/Os has been completed, analog and digital I/O requirements need to be specified. One consideration during this step is that analog sensors and actuators generally are more expensive than digital, so where possible it is more cost-effective to use digital I/O hardware.

The actual design of a DAQ system is an iterative process of comparing the specifications for candidate sensors/actuators and the DAQ hardware to be interfaced with these components. The format for data display must be chosen, as well as the preferred hardware or software to be incorporated into the DAQ system.

For each I/O channel, additional information is needed about the intended sampling rate, signal conditioning required, multiplexing, and range and resolution. Sections 6.2.1 through 6.2.4 explain these functions and how they relate to the selection process.

6.2.1 Sampling Rate

Sampling rate specifies how often the ADC takes place. A faster sampling rate acquires more points in a given time. High sampling rates generally are required for signals that vary quickly. This would not be expected to be the case for treatment plant monitoring. Input signals, however, must still be sampled at a fast enough rate to reproduce the varying analog signal. Data that are acquired at too slow a sampling rate may not accurately represent the "real world" event. An additional function that is tied to the sampling rate is low-pass filtering of the signal. Filtering will need to be performed just above the sampling frequency to prevent a distortion known as aliasing.

6.2.2 Signal Conditioning

Transducer outputs often must be conditioned to provide signals suitable for the digital input boards in the DAQ system. Signal conditioning accessories amplify low-level signals. They also isolate, filter, excite, and bridge complete transducers to produce high-level signals for the DAQ system.

The most common type of signal conditioning is amplification. Low-level signals are amplified to increase resolution and reduce noise. For highest accuracy, the signal should be amplified so that the maximum signal equals the maximum input range of the ADC.

Transducer signals may be isolated from the computer for safety purposes because the system being monitored may contain high-voltage transients that could damage the computer. Isolation also may be used to ensure that the readings from the plug-in board are not affected by differences in ground potentials or common-mode voltages. When the DAQ board input and the signal being acquired are each referenced to ground, problems occur if there is a potential difference between the two grounds. This difference may lead to a ground loop, which may cause an inaccurate representation of the acquired signal or may damage the measurement system. Using isolated signal-conditioning modules would ensure accurate acquisition of the signals. Filtering removes unwanted signals from the signal being measured. A low-pass filter is used on slowly varying signals, such as temperature, to attenuate higher-frequency signals that can reduce the measurement accuracy. Anti-aliasing filters are used to remove frequency components higher than the input bandwidth of the board.

Some sensors require external voltage or current excitation. Signal conditioners for such sensors such as ISEs or thermocouples usually provide the required excitation or reference voltages.

Many transducers have a nonlinear response to changes in the parameter being measured. Signal conditioning can linearize the transducer output, reducing this processing need in the DAQ system.

6.2.3 Multiplexing

Most ADC boards use one or two analog to digital (A/D) converters and connect these to the input channels sequentially. This method, known as continuous scanning, is significantly less expensive than having a separate amplifier and A/D converter for each input channel. The multiplexer selects and routes one input to the A/D converter for digitizing, then switches to the next input and repeats. The conversion reduces the effective sample rate of each individual channel in proportion to the number of channels sampled. External analog multiplexers also can be used to increase the numbers of channels a board can measure. Using this method, the effective sampling rate is reduced in proportion to the number of external multiplexers.

Because the multiplexer switches between channels, a slight time shift is generated between each channel sample. This is critical only for applications such as phase analysis of time-varying signals in which the time relationship between inputs is important. By scanning the input channels within microseconds, the effect of simultaneous scanning can be created if required.

6.2.4 Range and Resolution

Range refers to the minimum and maximum levels the A/D converter can span. DAQ boards offer selectable ranges so that the board is configurable to handle a variety of voltage levels. Selectable ranges allow matching of the signal range to that of the ADC for increased accuracy.

Resolution is the number of bits used by the ADC to represent the signal. The higher the resolution, the higher the number of divisions the voltage range is broken into and the smaller the detectable voltage change. For example, a 3-bit ADC divides the analog range into 2^3 or 8 divisions, whereas a 16-bit ADC divides the analog range into 2^{16}, or 65,536 divisions. Each division is represented by a unique binary code.

The range, resolution, and gain available on a DAQ board determine the smallest detectable change or ideal code width in voltage. The ideal code width is $V/(G*2R)$, where V is the voltage range, G is the gain, and R is the resolution in bits. For example, for a range of 10 volts, a gain of 100, and 16-bit resolution, the ideal code width is 1.5 μV.

6.3 Computer Hardware

Up to this point, the discussion has been focused on getting the signals into the DAQ computer, but no mention of the computer or software has been made. Many different computer bus structures are available for performing the DAQ task. A bus structure defines pathways for a collection of data addresses and control signals, and is used to interface a computer with its peripherals. Several currently used buses include the IBM PC/AT-compatible industry standard architecture (ISA) bus, the peripheral component interconnect (PCI) bus, the standard (STD) bus, and the VERSAmodule Eurocard (VME) bus. The PC/104 architecture is functionally the same as ISA and is popular where size is an issue. There also are nonbused-type systems such as PC Cards, also known as Personal Computer Memory Card International Association (PCMCIA) Cards.

In this brief summary the focus is on the PC (i.e., ISA) bus. This bus structure was chosen because many manufacturers support the PC bus, it has wide availability, and most vendors offer Microsoft® Windows®-based interfaces. PC-based DAQ systems have evolved from simple A/D devices into fully functioned instruments with sophisticated measurement capabilities, resulting from advances in higher speed buses, remote connectivity, and software with a graphic user interface (GUI).

A typical DAQ plug-in board includes several analog input channels multiplexed into a single instrumentation amplifier and A/D converter, which is clocked at rates around 200,000 samples per second. Many DAQ boards also include some combination of analog output channels for control and waveform generation, transistor-transistor logic (TTL)-compatible digital lines, and two or three counter/timers. Sampling rates higher than 750,000,000 samples per second now are being offered. The desired acquisition rate must consider the speed of the A/D converter (samples per second) as well as the bus speed (bytes per second) for continuous data collection.

The ability to move large amounts of data between a plug-in DAQ board and system RAM in a PC for real-time analysis and display is crucial to high-speed measurement applications. Compared with the legacy ISA bus, continuous data acquisition rates can be orders of magnitude faster for the newer buses such as PCI. Most bus systems, including ISA, have the option to use direct memory access (DMA) to move data from the peripheral DAQ board to the system RAM or disk storage. DMA improves speed because it bypasses the central processing unit (CPU) when transferring data.

Signal conditioning hardware may be used to preprocess the analog and digital signals connected to the DAQ board. In addition, external signal conditioning with electrical isolation is used when protection from high voltage or ground loops is required. Due to the rapid advancement in PC-based DAQ systems, the gap between plug-in DAQ boards and General Purpose Interface Bus (GPIB)-based instrumentation is narrowing in terms of resolution and sampling rate.

The growing popularity of notebooks with PC Cards has fueled the development of a wide range of PC Cards with DAQ functionality. New developments in DAQ technology pack the performance of a full-size plug-in DAQ board into the size of a small PC Card, so that notebooks with PC Cards can be used for very sophisticated DAQ applications. PCMCIA's Cardbus, a new standard that combines PCMCIA with PCI architecture, shares PCI's 32-bit operation, 132 Mbytes/second bandwidth, and bus-mastering features.

Both PCI and PCMCIA deliver true multiplatform solutions. PC Cards are available on most portable and desktop PCs, as well as in Apple PowerBook™ computers. PCI has become the standard backplane bus, being used in all Pentium®-based PCs, Apple Power Macintosh™ computers, and Digital Equipment Corporation (DEC) Alpha™ workstations.

PC software for DAQ continues to improve with the emergence of reliable 32-bit operating systems, Windows® 95/98, and Windows® NT. To make these systems work effectively with the DAQ hardware, DAQ driver software manages the operation of hardware and low-level system resources. The increasing sophistication of PC-based DAQ capabilities, combined with the increasing number of interface options (PCI, PCMCIA, RS232, RS485, etc.), has increased the importance of the DAQ driver software. A good driver software should provide (1) easy access to the full hardware functionality of the DAQ hardware and (2) an applications programming interface that is scalable across hardware and across several platforms. Depending on the DAQ hardware and software, disk operating system (DOS) handles some applications better than Windows®. DOS takes about half as much time as Windows® to process data, and hence is more useful for real-time applications.

Standalone measurement devices are specialized measurement devices, well suited for field applications requiring quick hookup close to the signal source and measurement of as many as 16 inputs per module. Multiple modules can be distributed and linked together as needed. The input signals connect to the module, which in turn is linked to the PC or network with a low-cost cable (typically RS485). Distributed modules allow more flexibility than ISA or PCI-type bus systems. These modules can take sensitive measurements, store and process the data, then transfer the data to the network or computer of choice. These modules have a DAQ rate of 50 kilohertz (kHz) and have local memory for 16K readings. They require direct current (DC) power and consume from 1 to 3 watts (W).

The DAQ system costs can range from $10 to $2,000 per channel, depending on the type of signal conditioning required and the speed involved in acquiring and transferring data. A typical system costs about $2,000 for hardware (excluding a notebook computer) and $1,000 for software. The system should handle a variety of signals and acquisition rates as high as 100 kHz. A typical Ethernet or high-speed system costs about $600 to $2,000. The various aspects/components of a DAQ system are depicted in Figure 6-1.

6.4 Driver Software

The majority of DAQ applications use driver software to directly program the various registers of the DAQ hardware, thereby managing its operation and integration with the computer resources such as processor interrupts, DMA, and memory. Driver software hides

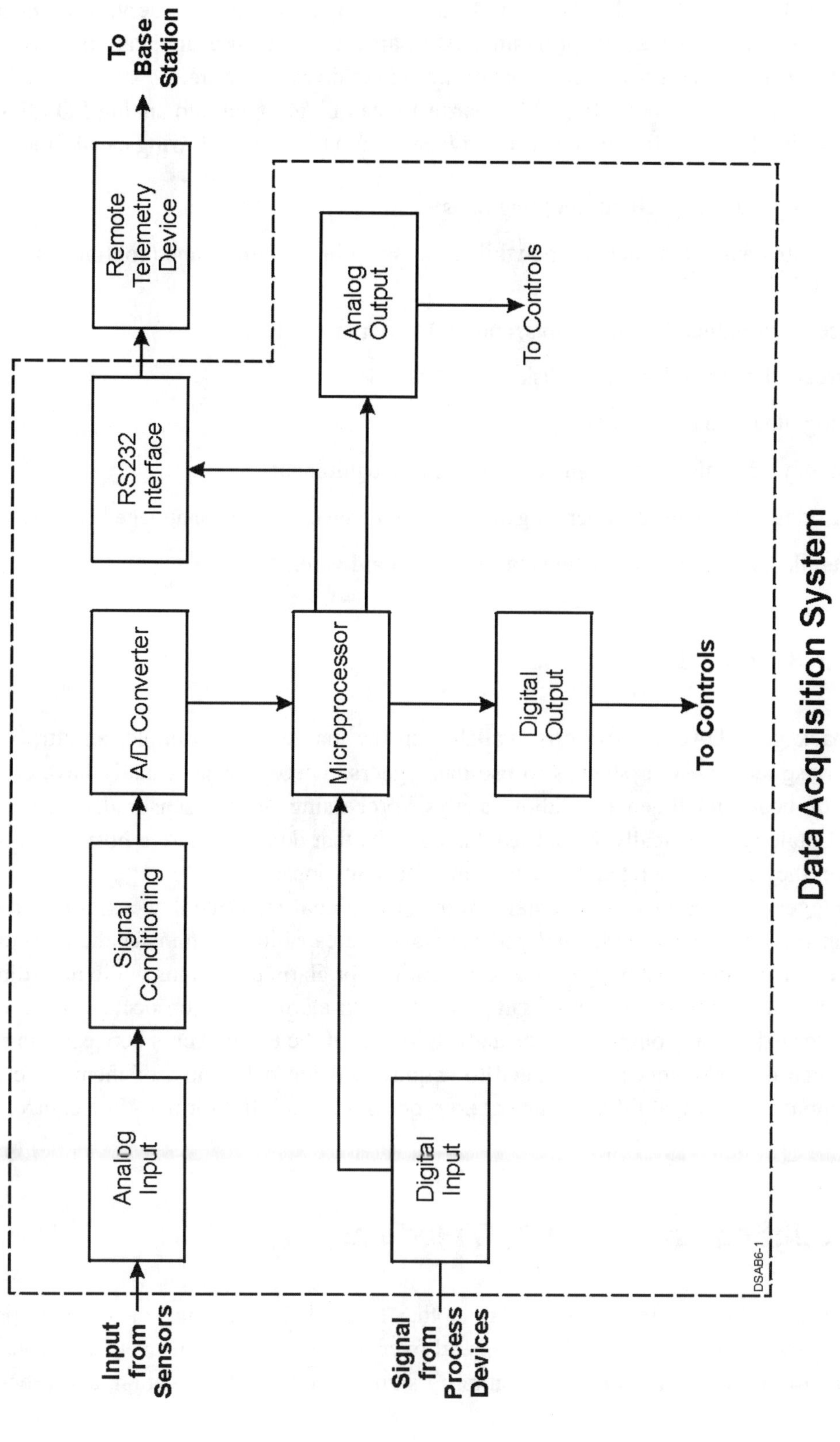

FIGURE 6-1. Data acquisition system hardware components.

the low-level, complicated details of hardware programming while preserving high performance. An efficient way to program DAQ hardware is to use application software, which adds analysis and presentation capabilities to the driver software.

Driver functions for controlling DAQ hardware can be grouped into analog I/O, digital I/O, and timing I/O. Full-function driver software should have the following capabilities:

- Acquire data at specified sampling rates
- Acquire data in the background while processing in the foreground (continuous DAQ)
- Use programmed I/O, interrupts, and DMA to transfer data
- Stream data to and from the disk
- Integrate more than one DAQ board
- Integrate seamlessly with signal-conditioning equipment
- Be compatible with the operating system and the programming language being used
- Handle all the critical functions of the DAQ hardware.

6.5 Dataloggers

"Prepackaged" DAQ hardware is available on the market. For example, an attractive way to construct a DAQ system is to use dataloggers. These devices usually have some form of on-board intelligence to allow analysis/processing of raw sensor data prior to storage. Dataloggers typically are self-contained units that do not require a host controller to operate; they can be operated unattended in almost any location.

Dataloggers have a distinct advantage over conventional interface devices in that they operate in this standalone mode, and yet they have the capability to transfer the data to a host system on request. Dataloggers also can monitor for alarm conditions while recording, for economy, a minimum number of samples. When an alarm condition occurs, the event will be recorded and any outputs will be activated, even if the event occurs between sample times. Because dataloggers are designed to acquire and log data, most of them have, at best, minimal output capability and almost no processing capability for real-time control of a plant.

6.6 Mounting Racks with I/O Modules

Mounting racks with I/O modules are another method for acquiring data that is particularly attractive to sites where I/O is spread over long distances; however, they can be used in any DAQ application and may simplify signal conditioning, isolation, and processor burden.

Mounting racks are convenient to use because the A/D I/O modules can be replaced easily and they provide straightforward integration of the sensors to the systems. As such, they are a good choice for DAQ and control in an RTS. Interfacing the RTS using RS232/RS485 connectivity between the controller and the remote monitor is one viable option.

The I/O mounting racks and modules, such as those manufactured by OPTO 22, have the convenience of modularity. The I/O modules with industry-standard sizes screw onto the racks. The sensor inputs are external to the modules and are connected onto the mounting racks. If a module fails, the sensor inputs do not have to be disconnected. This permits field replacement of defective modules by semiskilled technicians with minimal risk to the RTS system.

6.7 COTS Availability of Data Acquisition Systems

Specification data from suppliers of several DAQ systems are presented in Table 6-1. Additional information is provided in individual data sheets in Appendix D. Table 6-1 includes representative features of several DAQ systems, including DAQ boards, dataloggers, and mounting racks with I/O modules.

DAQ boards plug into PCs and typically require a terminal box to which leads from a sensor are connected. The use of DAQ boards requires a bus-based PC at the remote site. Any repair would necessitate replacing the entire DAQ board. The PC needs to be opened to replace the board, increasing the chance of inadvertent damage to the internal hardware of the computer.

Dataloggers are standalone systems that do not require a host computer at the remote site. They are capable of receiving analog inputs from sensors, and some have digital I/O ports for limited real-time control of various plant processes. Some of the vendors have considerable RTS experience. Dataloggers have limited real-time control capability. In case of channel failure, the entire unit may need to be replaced or removed and sent for repair.

Mounting racks with analog and digital I/O modules have standalone capability similar to dataloggers coupled with ease of replaceability of A/D signal-conditioning modules. These modules either snap or screw onto the mounting rack. If an individual module fails, it can be replaced easily. As an example, specific components and costs for the I/O mounting racks and modules manufactured by OPTO 22 are presented in Table 6-2.

Vendors or consultants can provide assistance in designing DAQ systems for an RTS at a water treatment plant. The key component in the design process is the list of monitoring and control functions associated with the specific plant, including monitors, pumps, pressure- and flow-monitoring devices, tank level sensors, utility meters, and any other function that demands monitoring for either regulatory or operational reasons. Then the power requirements, including types and ranges of inputs and outputs from each device, need to be identified. With this information, DAQ specialists can assist in tailoring a DAQ system to a specific plant at a cost-effective price.

TABLE 6-1. Representative Data Acquisition System Options

Manufacturer/ Model	Acromag AVME9510-1	Campbell Scientific, Inc. (CSI) CR23X	Data Electronics Datalogger DT500	Daytronic 10KU-EL	E. Clark & Associates ALMEMO 2290-8	Handar 555	In-Situ, Inc. Well Sentinel 3100
Principle of Operation	DAQ board interfaces digital and analog I/O signals to VME bus	Self-contained, rugged datalogger with nonvolatile data storage, battery backup, and real-time clock	Microprocessor-based, or main-powered devices designed to operate in any environment, independent of host computer	High-speed front-end data acquisition and logging	Automatic acquisition datalogger	Datalogger with DAQ board and modem/radio options	Four-channel datalogger; can read and store 500,000 readings
Interface with Telemetry Device	RS232	RS232 and CSI standard 9-pin	RS232 full duplex	RS232	RS232	SCI-12 Smart Sensor interface	RS232 serial port, upgradable to RS422/485; IBM 9-pin
Bits of Resolution	12-bit A/D, 32-μS conversion rate	15-bit	15-bit/1 μV; accuracy 0.15% of full scale	16-bit	16-bit A/D	14-bit	0.01% full scale; accuracy $\pm$0.1% full scale
Input Channels	16 differential or 32 single-ended high-level inputs; 2 output channels	12 differential or 24 single-ended analog inputs, 8 digital I/O ports	10 differential channels, 4 inputs per channel	160 analog inputs	4 analog inputs, 1 digital input	16 analog inputs, 8 digital inputs, 8 digital outputs	4 external, 2 internal
Cost	NR	$1,090	$2,575 with battery and software	$2,195	NR	NR	$3,800
Comments	Has both analog and digital I/O; problem with one channel may require replacement of whole board	PC208W datalogger support software for Windows®; also provides remote telemetry options (RF, phone, cellular, satellite)	Signals do not have to be conditioned; 1-MB storage space (340,000 readings) in PCMCIA card	Datalogger interfaces with any computer/ operating system and programming language; Labview software	Different sensor types can be connected without special programming	NEMA-4 enclosure houses 12-V battery, choice of telemetry or data storage options; experienced in remote telemetry; system has been used for air and water quality management	Nonvolatile 1024K memory; fits in 4-inch Schedule 80 PVC well casing, adaptable to 2-inch-diameter wells; 967K data storage; used to monitor water quality/process parameters

TABLE 6-1. (continued)

Manufacturer/ Model	ITT Barton DataScan DS40	Logic Beach, Inc. Minilogger	OMEGA OM-220	Onset Computer Corporation Tattletale 5F	OPTO 22 PB16AI	OPTO 22 PB16AH
Principle of Operation	Standalone datalogger with internal 2400-baud modem	Self-contained portable datalogging system with system microprocessor, data storage memory, A/D converter, and signal-conditioning circuitry for inputs and outputs	Remote-controlled portable datalogging system	Specialiy computer boards with integrated hardware and software components. Analog inputs from sensors, digital I/O, nonvolatile program and data storage options	Analog I/O mounting rack accommodates 16 I/O modules and interfaces with brain board via a 50-pin header connector	Digital I/O mounting rack accommodates 16 I/O modules and interfaces with brain board via a 50-pin card edge connector
Interface with Telemetry Device	RS232 full duplex; RS485, half-duplex; serial board	RS232	RS232, RS485	RS232	RS485	Six 32-screw terminals; 50-pin male card edge connector
Bits of Resolution	16-bit; ±0.1% of full scale for 4-20 mA inputs	12-bit	12-bit dual-slope integrating	12-bit	12-bit	12-bit
Input Channels	6 analog inputs, 2 analog outputs	24 analog, 1 digital	4 modules maximum, 4 channels per module; can select digital or analog modules	14 digital I/O, 8 analog I/O	16	16
Cost	NR	$1,200 for Minilogger, $1,800 for Modulogger	$1,350	$500	$1,000	$1,000
Comments	Used for DAQ and telemetry; NEMA-4/IP66 enclosure	Reprogramming of datalogger possible with remote PC; Hyperware software provided; has control functions; battery backs up SRAM (15,000 samples)	Inexpensive, versatile	Small, light, inexpensive; program stored in Flash EEPROM, embedded intelligence; used for remote monitoring in aerospace, oceanography, and several other applications; 480 kB RAM	Suitable I/O modules have to be selected; modules easily replaced, multiple source	Suitable I/O modules have to be selected; modules easily replaced, multiple source

EEPROM= electrically erasable, programmable, read-only memory.
SRAM = static random access memory.

TABLE 6-2. Typical OPTO 22 System Components and Estimated Costs

Item Description	Estimated Unit Cost ($)
Digital mounting rack	170
Communications interface; modem with built-in power supply	250
Brain board for digital mounting rack	221
DC output 5-V logic	10
AC output 5-V logic	10
AC input 5-V logic	10
DC input 5-V logic	10
Analog mounting rack	170
Brain board	225
0- to 5-V input module	55
4- to 20-mA input module.	45
0- to 5-V output module	95
4- to 20-mA output module	95

7

COTS Data Transmission Options

Data transmission alternatives are undergoing extensive development and advancements. The overview presented in this chapter is intended to provide the reader with the basic features associated with currently available technologies. As data transmission/remote telemetry alternatives become more sophisticated, the transmission of data is expected to become more reliable, easier to accomplish, and less expensive.

7.1 Overview of Data Transmission/Remote Telemetry Options

Telemetry is defined as the sensing and measuring of information at a remote site and then transmitting that information to a convenient or centralized location. An RTS needs to include data transmission so that automated process monitoring, equipment status, and process control can be accessed without actually having to travel to the facility.

Telecommunications peripherals allow the interrogation of a DAQ system from a remote computer. The RTS may be land-based phone lines, cellular phone systems, radiofrequency (RF) links, or satellite communications. The block diagram in Figure 7-1 identifies the major features of these four remote telemetry options. The plain old telephone system (POTS) or land-based phone line depends on a physical connection between the point of transmission and the point of reception, as well as the telephone switching stations at both points. Cellular phones generally transmit the data (via an RF link) to a cellular base station, which then sends the data to the final destination through telephone systems using an appropriate medium. Cellular phones are applicable only when they are within range of the

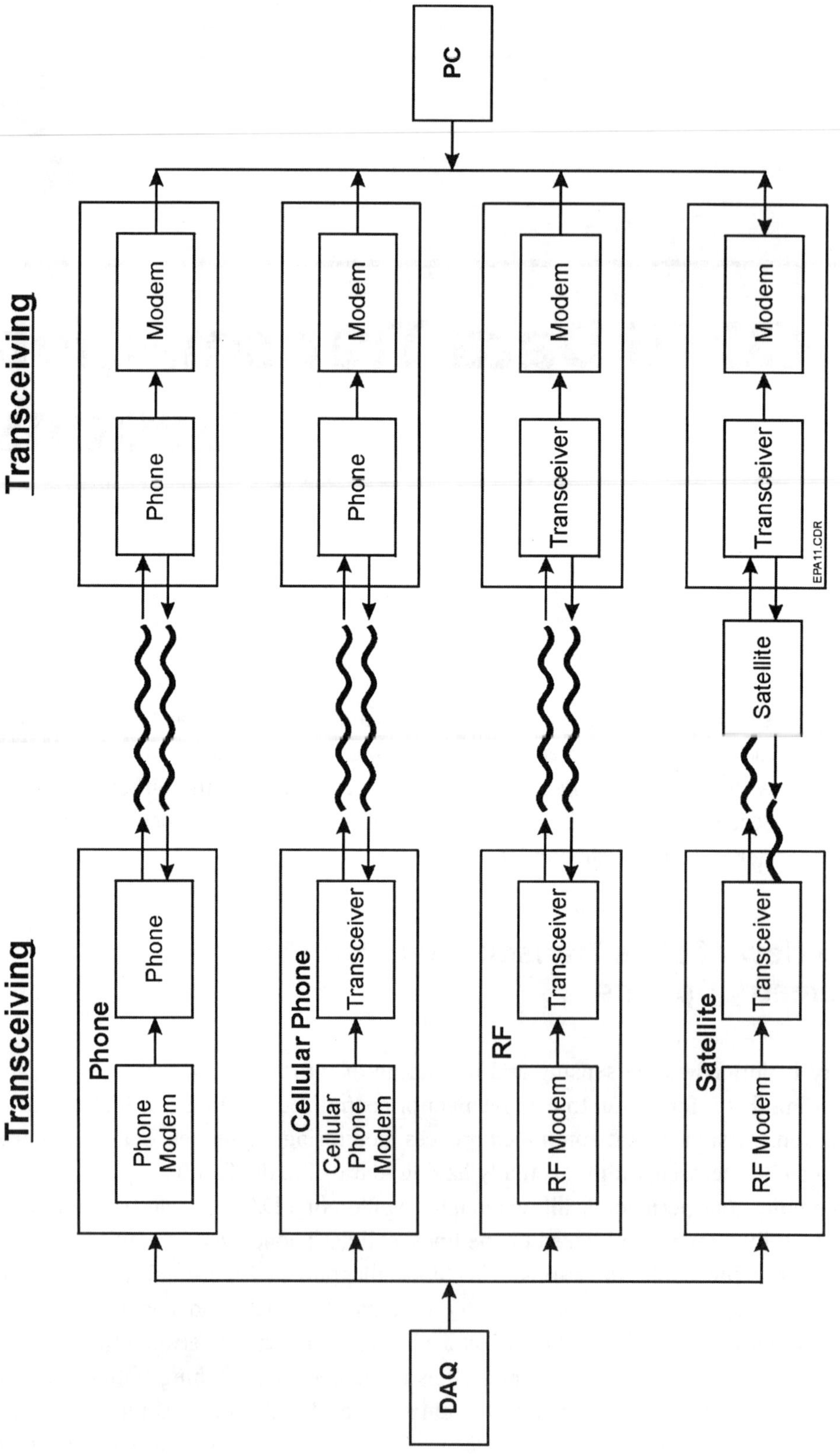

FIGURE 7-1. Remote telemetry block diagram.

cellular network. The physical link between the telephone switching stations can be, and most often is, through either an RF link or a satellite. RF links depend on transmission from point A to point B by radio waves. One problem with using radio waves is that their transmitting range can be limited. Because radio waves travel in a straight line and do not follow the curvature of the earth very well, the transmitter and receiver must be within the line of sight, thus limiting the transmission range. To achieve longer transmission distances, taller antennas with greater power must be used. Satellite systems, which are a specialized RF link, overcome most of the problems inherent in POTS, cellular, and ordinary RF systems. With a satellite system, the signal is transmitted up to a satellite, then back down to a land-based station for final transmission to its destination.

The four remote telemetry options are discussed in Sections 7.1.1 through 7.1.4. Table 7-1 summarizes the advantages and limitations of the four technologies.

TABLE 7-1. Advantages and Limitations of Data Transmission/Remote Telemetry Options

Technology	Principle of Operation	Advantages	Limitations
Telephone	Data transmission through telephone lines	Hardware readily available Infrastructure in place Upgraded by the carrier Can be used for duplex communication at high rates Low cost per call Extensive range	Requires access to telephone line
Cellular	Data transmission through cellular network	Infrastructure in place Upgrades to system funded by carrier Does not require FCC-assigned RF Wide range as long as cellular phone is within range of a network	Needs cellular-compatible modems, which are not as readily available as telephone modems Digital data may be corrupted when used with mobile applications Cost per call is higher than for land-based telephones
Radio-frequency	Transmission by radio waves at MHz-GHz range	Fast response time No transmission lines needed Can be located in remote areas where it is not possible to use wire or coaxial cable Per call cost lower than for cellular phone system	Low range Needs FCC license (time-consuming) Requires line-of-sight May have interference from extraneous electrical noise Startup and maintenance costs higher User bears cost for upgrades to system
Satellite	Data transmission to and from satellites using RF technology	Extensive range Several options available Increasingly being used by large corporations to transmit data Cost of updating technology spread across the market Startup and per call costs expected to be the same as those for cellular technology in 2 to 5 years	Startup and per call costs higher than for cellular system

FCC = Federal Communications Commission.
GHz = gigahertz; MHz= megahertz.

7.1.1 Hard-Wired Land-Based Telephones

POTS communications require physical access to a telephone line and can be used for duplex communication transmission at reasonably high rates. Hardware is readily available and easy to interface to DAQ/control systems. Telephone communication links require a modem at both the transmitting and receiving site to convert the digital data into an analog format that the POTS system can transmit (see Figure 7-2). Because POTS lines are strung for many miles above ground, they are susceptible to lightning-induced voltage spikes. Thus, surge protection is highly recommended.

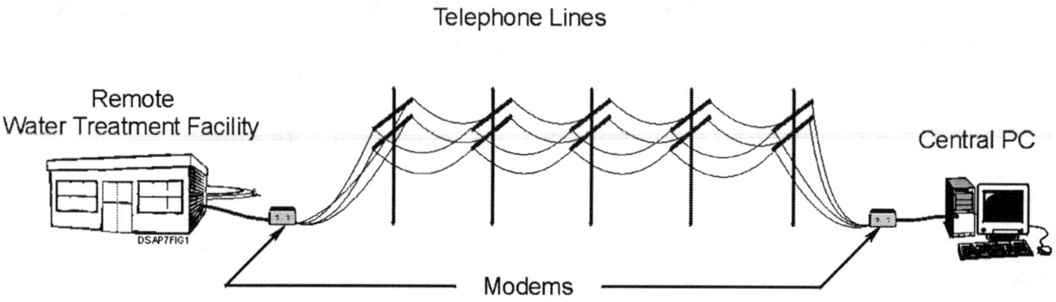

FIGURE 7-2. Data transmission via a hard-wired land-based telephone system.

The RTS can be connected to a telephone line, and the data can be transmitted to a required location at transmission rates of 50,000 bits/second (bps). Because the telephone hardware infrastructure is in place, the access cost generally is low, and the system typically is being upgraded by the carrier. Therefore, POTS is an easily implemented option, when available. Additionally, with the wide use of modems on personal computers, replacement parts are almost universally obtainable if a component should fail.

7.1.2 Cellular Phones

The front end of a cellular phone system is RF-based and requires that a cellular phone be within a service provider's coverage area. Cellular phones use a transceiver and a modem to transfer data between a DAQ system and a base station computer (see Figure 7-3). Cellular communications would require a subscription to the cellular network coverage at the RTS site. The cellular option is a convenient alternative to mobile applications or for locations where hard-wired telephone lines do not exist. Cellular communications require no license for operation by the user, because the Federal Communications Commission (FCC)-assigned frequency has already been licensed to the cellular provider.

For mobile applications, cellular communications can at times be slow, because digital data may be corrupted when the phone cell changes within the network. However, because an RTS application is a fixed platform, transmission speed should not be an issue.

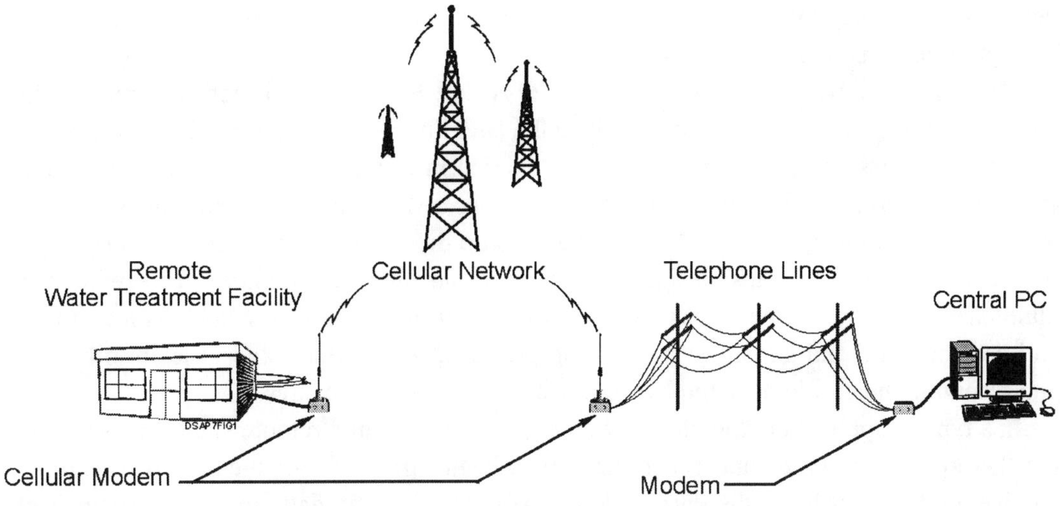

FIGURE 7-3. Data transmission via a cellular phone system.

The cost per call generally is somewhat higher than for a land-based telephone system. The infrastructure for cellular phone operation is already in place, and upgrades to the system undoubtedly will occur. Any upgrade costs will be absorbed, at least initially, by the carrier. A cellular phone system for data transmission can be cost-effective if direct land-based telephone service is not available.

7.1.3 RF Communication

RF communication systems use a radio modem and a low-powered transceiver at the remote station, and a transceiver connected to an RF base station at the computer site (see Figure 7-4). Several stations can be interrogated over a single ultrahigh frequency (UHF) or

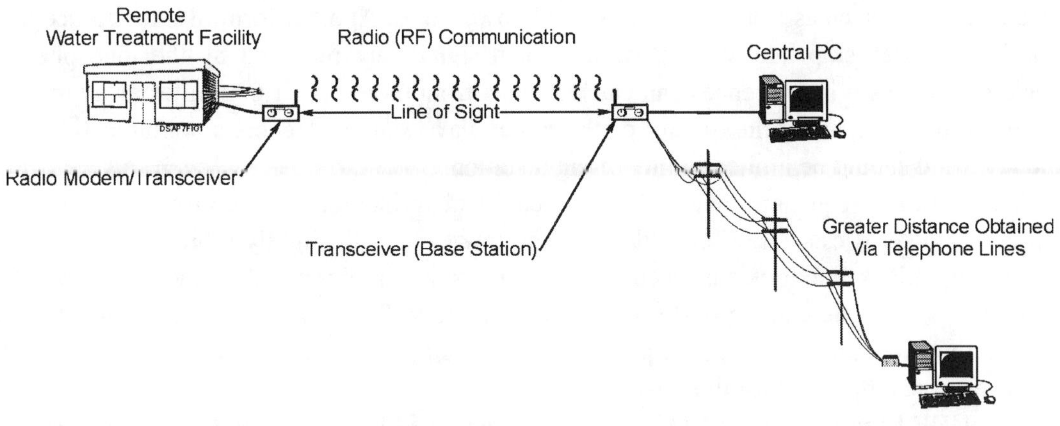

FIGURE 7-4. Data transmission via radiofrequency system communication.

very high frequency (VHF) system. Any station can serve as a repeater to extend the line-of-sight transmission of the base station.

The primary advantage of RF over POTS is that RF can be located in remote areas where it is not economically feasible to install a land line. Also, because radio waves do not employ transmission lines, there is a great reduction in concern over damage from inclement weather. Ease of relocation and operability over a wide range of conditions also make RF communications attractive. Properly designed radio links can provide low-cost, effective, and flexible data transmission systems that can operate for many years with little maintenance. Radio telemetry has been used for a variety of purposes, such as monitoring and control of oil and gas lines; monitoring pressure, temperature, and gases in oil wells; and receiving meteorological and hydrological data from remote sites.

In a typical application, the base station requests data from a remote site by transmitting a wake-up signal to send data. When the remote begins transmitting, the base reverts to the receive mode and collects the data package. After transmitting data, the remote goes back to the receive mode and awaits instructions from the base. The output of the sensors at the remote site usually is converted to digital data by the on-site DAQ system. This signal is delivered to a modem that converts it to an analog form that can be frequency-modulated to the RF carrier. This signal typically is in the range of 300 to 3000 Hz. When the base receives the analog data, the base modem converts it back to digital data. If data are to be sent only one way, a standalone transmitter is required at the transmitting site and a standalone receiver is required at the receiving site. If data are to be sent both ways, a transceiver is required at both locations. Commercial RF modems are available and suitable for use with DAQ/control systems.

RF telemetry systems are normally configured as a fixed-base station that obtains information from another fixed station at a remote site. The FCC has allocated certain frequencies that can be used for fixed operation. Certain frequencies are available for RF transmission in the low band (25 to 50 MHz), mid-band (72 to 76 MHz), VHF band (150 to 174 MHz), UHF band (450 to 512 MHz), and 900 MHz (928 to 960 MHz).

The low band from 25 to 50 MHz provides the greatest communication range because the path loss is less than at higher frequencies. This band, however, is subject to extraneous electrical noise, such as from automobiles and power lines. Another form of interference in this band, called skip interference, occurs when signals are reflected by the ionosphere back to earth. Skip interferences can cause signals from more than 1,000 miles away to be received, thus degrading the quality of the direct wave signals. Hence, communication in the low band should be limited to line-of-sight stations.

There are 62 frequencies available for fixed telemetry stations in the mid-band. Propagation characteristics are similar to those of the low band with slightly more path loss and less susceptibility to extraneous electrical noise and skip interference. Extraneous electrical noise and skip interference cause fewer problems at the VHF frequencies (150 to 174 MHz) than at the low- and mid-band frequencies. The range of transmission in the VHF band, however, is less than that for the lower bands.

The UHF band (450 to 512 MHz) is most often used because of the large number of channels available. The transmission range is not as good as in the VHF band, but the UHF band is free of most extraneous electrical noise and skip interference. Absorption by trees and foliage causes a greater path loss, but penetration into buildings is better because the short wavelength signal has the ability to reflect off conducting objects.

Because the RF spectrum is finite in size, its use is regulated by the FCC, and transmission on most frequencies requires the system to be licensed and monitored during operation. Obtaining a license for exclusive use of a frequency can be an expensive and time-consuming endeavor. The startup and maintenance costs generally are higher, whereas per call costs typically are lower compared to costs for a cellular phone system. The user is the sole owner and operator of the infrastructure and bears the full cost for upgrades as technology advances.

7.1.4 Satellite Telecommunication

Satellites offer another telecommunication alternative for field stations where phone lines do not exist or RF systems would be impractical. A satellite performs the role of a relay station. Signals are transmitted from the remote site to a satellite, back to a stationary receiving station, and finally via telephone line or RF to the ultimate destination (see Figure 7-5). Satellite systems are already in place with many more to follow in the next few years. The startup and per call costs generally are more expensive than costs for existing POTS lines or cellular phones, but satellite use is expected to become an increasingly cost-competitive option. The various types of satellites are low earth orbit (LEO), medium earth

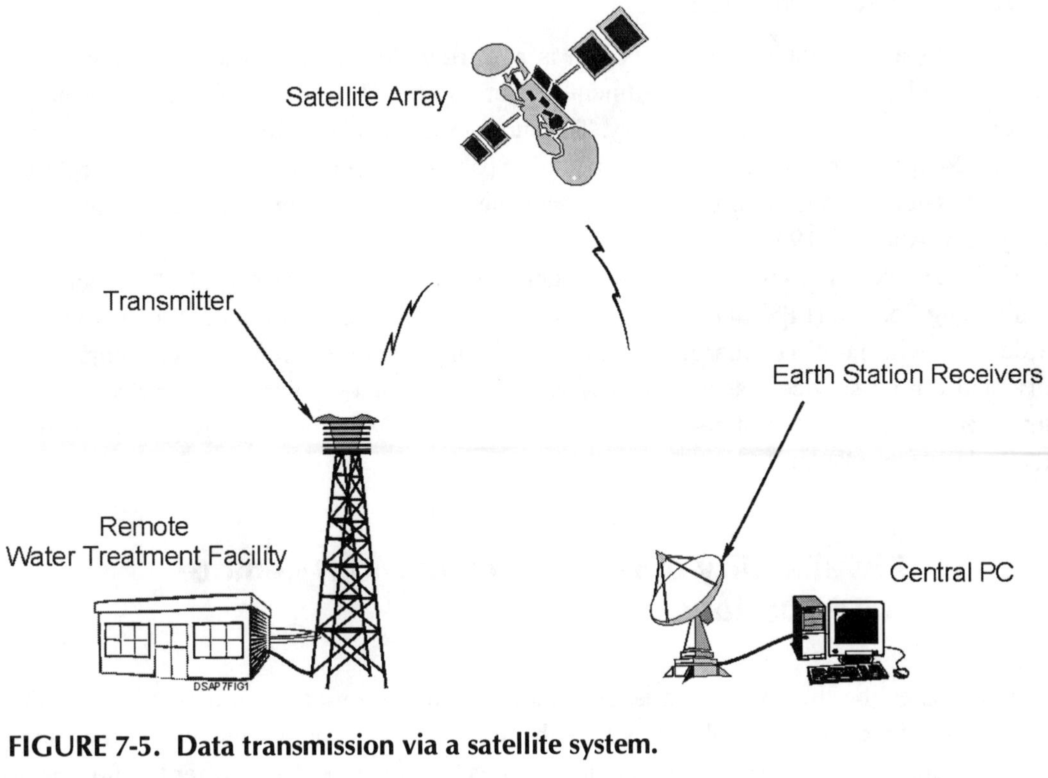

FIGURE 7-5. Data transmission via a satellite system.

orbit (MEO), and geostationary satellites. The satellite distance from the earth is a mile or more for LEO satellites to several thousand miles for geostationary satellites. Geostationary satellites orbit at the same speed as the earth rotates. They cover vast areas of the earth. As few as three satellites placed over the equator can cover the entire earth. On the other hand, LEO satellites have a smaller range, requiring up to 30 for full coverage of the earth.

7.1.4.1 Commercial Satellites

Commercial communications satellites transmit over two sets of radiofrequencies: C-band, which offers lower transmission power over a wide geographical area and requires a large reception dish; and Ku-band, which offers higher transmission power over smaller geographical areas and can use a smaller reception dish. Examples of companies that provide satellite services are SATCOM and PanAmSat. The Honeywell/Racal satellite communications system gives access to telephone and fax service aboard an aircraft. It also allows direct pilot-to-controller voice contact and a datalink for operational and engineering data communications.

Satellites are already in use by corporations to set up company intranets, to track freight movement, and to monitor data acquired in remote places. Because commercial satellite transmission is a rapidly emerging technology and has the potential for widespread use, the cost of keeping current will be spread across the market instead of being borne by a single user. Other examples of currently available commercial satellites are the Geostationary Operational Environmental Satellite (GOES), ARGOS, INMARSAT C, and Meteosat.

7.1.4.2 Military Satellites

The Global Positioning System (GPS) is a worldwide, satellite-based radionavigation system developed by the U.S. Department of Defense (DoD) to be DoD's main radionavigation system well into the next century. The current system, also called a constellation, consists of 24 operational satellites. The Air Force Space Command (AFSC) formally declared the GPS satellite constellation as having met the requirements for full operational capability as of April 27, 1995.

GPS provides two levels of service: Standard Positioning Service (SPS) and Precise Positioning Service (PPS). The SPS is available to all GPS users on a continuous, worldwide basis with no direct charge. The PPS is a highly accurate military positioning, velocity, and timing service that will be available on a continuous, worldwide basis to users authorized by the United States. PPS was designed mainly for military use, although it is available to other Federal Government users.

7.2 COTS Availability of Data Transmission/Remote Telemetry Options

When selecting the data transmission/remote telemetry option for a particular site, the remoteness of the location probably will be the primary driver for system cost. As developments in telemetry continue to evolve, more companies are expected to offer the full range of options from POTS to satellites. Correspondingly, prices for these services are expected to continue to decrease as competition increases in this market.

Table 7-2 provides information on features of representative COTS data transmission/remote telemetry options. Appendix E includes the individual data sheets with specifications on the data transmission/remote telemetry options from various vendors coded to their position on Table 7-2. Sections 7.2.1 through 7.2.4 describe the options identified for each technology.

7.2.1 POTS Telemetry

When a POTS line is available, POTS telemetry is an easily employed option. A POTS line with an external modem is one suggested configuration, because this can provide better protection of an RTS system in the event of a direct lightning strike on the POTS line. Some available modems are presented in Table 7-3.

7.2.2 Cellular Phone Telemetry

For places where there are no POTS lines nearby, cellular phones provide a readily available telemetry medium. The phone modems presented in Table 7-3 may or may not be compatible with cellular phones operating at higher data rates. Representative systems from

TABLE 7-2. Features of COTS Data Transmission/Remote Telemetry Options

Manufacturer/ Model	Synetcom RTU-1	ITT Barton DS40	Dancer Communications, Inc. Messenger 500 Series	Johnson Data Telemetry DL-3412/ DL 3422	Adcon Telemetry μT
Technology	RF transmission, 2400 baud; spread spectrum, wireline, and other options soon	Internal modem, 2400 baud	Built-in modem; voice and data communication capabilities	RF transmission	RF transmission
Range/ Sensitivity	RTU-1 datalogger with telemetry; 450 and 900 MHz transmission range	DS40	Wide range for phone or cellular	10 to 20 miles	1 mile
Cost	$750	NR	NR	$645	$1,200
Comments	Has cellular-compatible modem; range depends on transmission technology	Used in DAQ and telemetry; range depends on transmission technology	DAQ and remote telemetry	SCADA, water management, data acquisition	Asynchronous serial communication 19.2 kbaud; 418 MHz, 902-928 MHz

TABLE 7-2. (continued)

Manufacturer/ Model	Adcon Telemetry A730MD	InterOcean Systems, Inc.	duTec Remote Telemetry Unit (RTU)	Campbell Scientific, Inc. (CSI)
Technology	RF transmission	RF transmission, fiber optics, cellular, satellite	RF transmission; spread-spectrum technology for distances of less than 20 miles	Radiofrequency (R) Telephone (T) Cellular (C) Satellite (S)
Range/ Sensitivity	20 km	NR	RTU	Range depends on transmission technology
Cost	NR	$15,000 for RF, $25,000 for satellite	$1,200	$2,272 (R) $1,425 (T) $2,455 (C) $3,200 (S)
Comments	Programs stored in internal EEPROM	Versatile system integrator; does not manufacture components; range depends on transmission technology	RTU calls host whenever user-defined event occurs	The DAQ systems can be interfaced with various telemetry options

NR = not reported.
RTU = remote telemetry unit.
SCADA = supervisory control and data acquisition.

TABLE 7-3. Modems Used for Land-Based Phone Telemetry

Vendor	Model	Description	Cost ($)
U.S. Robotics	Sportster 336E	36.6 kbps data	219
Boca Research	M144EW	14.4 kbps data	125
Zoom	Z2836	33.6 kbps data	199

kbps = kilobits per second.

Transtel Group, Inc., Synetcom Inc., and most cellular service providers are fully compatible at higher data rates. One limitation for utilizing a cellular system is that transmission components may not always be available in truly remote areas.

7.2.3 Radiofrequency Telemetry

For areas where land-based phone lines are not available and which are out of the range of cellular networks, RF telemetry can be a viable option. RF transmissions can be achieved across various bands. The UHF band is most often used because of the large number of channels available. Although the UHF range is not as good as the VHF range, the UHF band is free of extraneous electrical noise. One example of an RF system is the Model DL-3412 from Johnson Data Telemetry Corporation with an internal RNET modem transmitting in UHF. The transmission range using directional antennae is expected to be 10 to

20 miles, depending on the terrain. The use of repeater sites can increase the range of transmission but would also increase cost. The interface requirements for this type of system would include an RS232 interface and a DAQ system that would be able to deliver "request to send"/"clear to send" signals.

7.2.4 Satellite Telemetry

Satellite telemetry significantly increases the data transmission range. The startup and per call costs, however, are more expensive than for POTS or cellular phones. Because the number of providers in the satellite telemetry service is increasing daily, satellite telemetry is expected to become an economical option in the near future.

The ARGOS satellite is managed by the National Oceanic and Atmospheric Administration (NOAA) and the U.S. Department of Commerce. Its main purpose is to support environmental monitoring. The ARGOS satellite system has three satellites in polar orbit at 90 degrees with respect to each other; therefore, there are periods when no satellites are visible from a fixed point on the earth. They orbit at 500 km from the earth and require low transmission power. Because water quality monitoring is a regulated activity, RTS applications may be approved to use the ARGOS for telemetry transmission. Communication is only one way from the remote site to the base station via the satellites. The control data from the base station to the RTS site may need to be sent via one of the many commercial Internet providers that currently are providing transmission service. These systems use free bandwidth within the direct digital system (DDS) to provide telemetry via a small receiving antenna.

8

Selection Criteria

This chapter describes the criteria that could be used when selecting specific devices/components for incorporation into an RTS for monitoring and control of a small drinking water facility. Criteria are presented for selecting water quality parameter monitoring devices, process monitoring and control components, DAQ systems, and remote telemetry systems to transmit and receive data.

To expedite the selection process, the first activity should be to prepare a list of features associated with the particular water treatment facility. The water quality parameters and types of monitoring that are key to the specific operation should be identified. For example, residual chlorine should be monitored to verify disinfection operations. Next, the regulatory compliance requirements should be reviewed and considered when reviewing manufacturers' specifications and discussing the applicability of their device or system. Finally, the monitoring device is selected.

It is important to document the pump characteristics, operational functions, and the physical dimensions of the treatment system before selecting components to monitor flowrates, pressure, electrical usage, or other process functions. Facility specifications should be known or easily accessed when contacting technical representatives of process monitoring/control devices. Selection of components must be based on the plant's operational characteristics.

Data acquisition and telemetry components can be purchased as package items, but again, an inventory of what devices will be used on line and what options exist for transmitting data in the locality of the treatment plant should be identified before contacting suppliers.

After the specific requirements of the RTS have been identified, many feasible options may be found. Sections 8.1 through 8.4 provide examples of the process of selecting components suitable for monitoring water quality, process monitoring and control, data acquisition, and telemetry. Any references to specific types of products should not be construed as an endorsement or recommendation for their use.

8.1 Water Quality Monitoring and Control Devices

8.1.1 Selection Criteria

When selecting a specific water quality monitoring and control device, the following features should be considered:

- **Sensitivity.** Does the sensor have the sensitivity to measure the parameter at the concentration specified as the lower limit of the regulatory requirements?
- **Reliability.** Do the literature and previous users of the sensor indicate that the device can be expected to supply stable and reliable performance over extended periods of time?
- **Maintenance Requirements.** Does the sensor require extensive maintenance? Are there consumables, such as probe solutions, that demand frequent manual replenishment?
- **Online Design.** Does the sensor design lend itself to being incorporated into a pathway that would permit online calibration or remote validation of sensor performance?
- **Amenability/Compatibility.** Does the sensor have power demands and signal output compatible with other components so that it can be easily integrated into the RTS with minimal custom electronic components?
- **Cost.** Is the cost of the sensor such that the overall expense of the RTS makes it a viable option to manual site monitoring of the water facility?

8.1.2 Selecting Devices/Components

This section provides guidelines for selecting specific sensors, probes, and monitoring devices that may be included in an RTS for a small drinking water facility.

8.1.2.1 Selecting Microbial Contaminant Monitors

No COTS system currently appears to be available for online monitoring of microbial populations in water. The *CT* approach discussed in Sections 2.2.1, 2.5.1, and 3.1.1 may become viable for inclusion in an RTS for demonstrating regulatory compliance when the technology is commercialized. Particle counters could be used to indicate if microbial-sized units are present in potable water, but definitive confirmation would still have to be performed off line in a laboratory.

8.1.2.2 Selecting Turbidity Sensors

A major consideration when selecting a turbidity monitor is whether the device is intended for measurements over the expected operational range of ~0.5 to 100 NTUs. If a monitor reportedly functions over that range, it may not have the sensitivity and accuracy, i.e., ±0.1 NTU with repeatability of ±0.05 NTU, to consistently detect low levels of suspended materials in postfiltration water samples. Another consideration is if the system's power requirements (either low-voltage direct current [DC] or single-phase alternating current [AC] power) and signal output (either millivolt or 4 to 20 mA) are reasonably compatible with field utilities and RTS DAQ components.

One system with the above features is the optical technology that uses an incandescent light source to detect turbidity via scatter. This technology is available in several process monitors identified in Section 4.2. Care should be taken when judging a monitor's sensitivity. Turbidity monitors with detection levels of 0.30 NTU may be necessary for indicating the presence of microbial contaminants. Prior to purchasing a unit, confirmation of the system's detection limit should be discussed thoroughly with the manufacturer. Furthermore, because turbidity process monitors cost from $2,000 to $3,000, it is advisable to request references of previous purchasers and contact them for first-hand experience on the suitability of a system.

8.1.2.3 Selecting pH Sensors

Probe-type pH sensors have been available for many years. Most units cover the pH range from 0 to 14 pH units with sensitivity of <0.005 pH unit and signal stability of ±0.03 pH unit, noncumulative, per 24 hours. It is advisable to select an automatically temperature-compensated unit that performs over the working range that can be expected at the treatment facility. As discussed in Section 4.3, power supply typically is not a concern, because the probe generates either a millivolt or a milliamp output signal by a chemical process that is not driven by a power supply. However, line power would be required to operate any on-site meter for real-time readings. Probes typically cost from $500 to $800. The probe may require an analyzer with electronics that could cost more than the probe itself.

8.1.2.4 Selecting Chemical Dose/Residual Chlorine Monitors

Residual chlorine process monitors identified in Section 4.4 for possible RTS inclusion employ either colorimetric or electrode technologies. Both types of monitors reportedly can detect free chlorine to hundredths of a milligram per liter with sensitivity to a thousandth of a milligram per liter. A range of operation from 0 to 2.0 mg/L most likely would be used for drinking water monitoring applications. Linearity of response would be expected to be in the range of 0.5% with repeatability of 0.05 mg/L. The colorimetric monitors generally have consumables requiring replenishment, typically on a monthly basis. The electrode type of monitor requires maintenance, which may include membrane and electrolyte replacement as the cell performance degrades. Power requirements typically are line power with signal outputs as either millivolts or milliamps. Because these devices cost in the range of $2,000, discussions with the manufacturer and other operators who have used the monitors are recommended prior to committing funds for their purchase.

Care should be taken when selecting a probe to monitor free chlorine that is already integrated into a suite of sensors that rely on other sensor outputs. Under this scenario, very precise measurements of free chlorine can be achieved, but a domino effect also is possible. If one sensor fails to operate properly, its failure may cause other sensors to supply invalid measurements.

8.1.2.5 Selecting Temperature Monitors

Industrial bimetallic thermocouples were identified in Section 4.5 that operate over a range of −50 to 50°C with accuracy of ±0.1°C and are readily available for inclusion into an RTS. Low-voltage DC power typically is required for the meters used for these sensors, which have linear 4 to 20 mA or millivolt outputs to choose from. The cost of the meter also must be considered when selecting a temperature monitor. Options of insertion versus surface patch thermocouples have implications for ease of installation. Thermocouples offer a proven temperature-measuring technology with a wide range of applications and sensitivity, thus providing a viable option for inclusion into an RTS.

8.1.2.6 Selecting Total Organic Carbon Monitors

Based on the review of COTS TOC-monitoring devices in Section 4.6, the monitoring of this parameter at small treatment facilities would be a high-cost operation. The two process monitors identified, which typically are used in industrial or laboratory applications, each cost approximately $20,000. Although they reportedly are fully automated, they could require a level of attention that may not be available at most small treatment facilities. These factors need to be considered before selecting a TOC monitor for an RTS.

8.1.2.7 Selecting Trace Metals Sensors

Although several iron and manganese monitoring systems were identified (see Section 4.7), their costs, relative complexity, and typical industrial and laboratory applications have to be taken into consideration before selecting one of these devices for incorporation into an RTS.

8.1.2.8 Selecting Anion/Cation Monitors

The fluoride probes that were reviewed have a reported sensitivity of 0.02 mg/L and resolution of 0.001 mg/L. Accuracy of ±0.5% reportedly is achievable. Power supply of low-voltage DC with millivolt outputs is typical. ATC should be considered when selecting an ISE. A combination probe that includes both reference and measurement components costs about $300 to $500, with electronics for meters and outputs potentially costing an additional $1,000 to $1,500. Considerations of possible interferences, the effects of variable pH, delivery of buffer solutions, stability of response over time, and delivery of standard solutions to confirm operational integrity make the use of an ISE for remote fluoride monitoring a challenge beyond simply inserting a probe into a sampling stream.

Nitrate monitoring by an ISE has the same considerations. A process monitor that utilizes the electrode technology is available. The monitor includes online calibration and temperature controls. However, its cost of $6,200 may be a factor when selecting monitors for use in an RTS.

Several ISEs that are available for the detection of fluoride and nitrate are identified in Section 4.8. ISE process monitoring developments are likely to occur that should make a COTS option viable for future RTS applications.

8.1.2.9 Selecting Color Monitors

Although color-monitoring devices exist for industrial applications, as identified in Section 4.9, low-level color monitors specifically for drinking water were not identified during the literature searches that went into the database for this book. If monitoring devices do become available, the general criteria of required sensitivity, reported reliability, maintenance requirements, online design, amenability/compatibility with power and output requirements, and cost should be considered.

8.1.2.10 Selecting Hardness Monitors

ISE monitoring for calcium carbonate in water can be used to indicate hardness. Typical concentrations of calcium carbonate in potable water supplies range from 0 to 150 mg/L, although the level may be much higher if water is being drawn from a geologic formation that fortifies carbonates. Therefore, an electrode should be selected that will operate over the historical range of calcium carbonate in the water source. The same rationale for monitoring with an ISE for anions could be applied for hardness. It is important to identify any possible interferences, such as copper, zinc, nickel, or iron, that could restrict the use of a particular type of electrode, and to discuss with the manufacturer how best to validate the ISE's performance.

Online hardness monitors based on ISE response are available. One unit identified in Section 4.10 reportedly requires only 20 minutes of maintenance per month with a projected minimum electrode life of 6 months of full-time operation. With a unit cost of about $4,200 and consumables costs exceeding $1,000 annually, hardness monitoring may need to be a high priority to warrant incorporation into an RTS.

8.1.2.11 Selecting Total Dissolved Solids/Conductivity Sensors

To monitor TDS concentrations ranging from 0 to 500 mg/L or higher, a multirange sensor or several specific sensors with discrete ranges must be used. Sensor/meter combinations that offer multiple TDS range capabilities are identified in Section 4.11 that would offer a viable option. Capabilities of the sensor should include temperature compensation, calibration inputs for each range, and probe insertion into a process flow.

8.1.2.12 Selecting Particle-Counting Monitors

Because off-the-shelf microbial monitors presently are not a commercial option, particle counters and turbidity monitors have been widely used to indicate the presence of microbial-size particles in water. When selecting a particle counter, the ability of the system to detect particles of low micron dimensions should be considered. The counter should be able to differentiate particle sizes over a fairly broad range, from 2 to 750 microns. Because flow-rate fluctuations on some systems may bias results, the unit chosen must have constant-flow capabilities. Requesting references and talking to operators who have been using a specific unit may be warranted before making a purchase, because the monitors identified in Section 4.12 cost from $4,000 to $10,000.

8.1.2.13 Selecting Multiparameter Probes

The multiparameter probes identified in Section 4.13 provide an appealing option to solving multiple monitoring needs by identifying the parameters required and having them integrated into a single sampling head or flow cell by the manufacturer. Although multi-parameter monitoring is possible, the operational and performance characteristics of each sensing element within the multiparameter monitor should be reviewed to confirm that each has the sensitivity, reliability, ease of use, and data validation characteristics required for the specific application. Consideration also should be given to any interactions between probes and how they may affect data quality. For example, if a pH probe is interfaced with a residual chlorine probe, the facility could experience chlorine false-alarm situations due to errant pH probe performance. Each sensor in a multiparameter probe system can cost more than a single probe unit, but that generally is due to the advanced electronics used to interface the various monitoring functions.

8.2 Process Monitoring and Control Devices

8.2.1 Selection Criteria

When selecting process monitoring and control devices/components, the following features should be considered:

- **Range.** Does the sensor/monitor operate over the expected range for the parameter of interest?
- **Ambient Conditions.** Can the sensor/monitor operate under the expected ambient conditions of the water facility?
- **Amenability/Compatibility.** Does the sensor/monitor have power needs and signal outputs that are compatible with those of the other components?
- **Durability.** Is the component durable, requiring minimal maintenance?
- **Cost.** Is the cost of the sensor/monitor such that its expense will not significantly contribute to the overall cost of the RTS?

8.2.2 Selecting Process Parameter Devices/Components

This section provides guidelines for selecting process parameter components that may be included in an RTS for a small drinking water facility.

8.2.2.1 Selecting Flowrate Sensors

To monitor flowrate, the mechanical and nonmechanical devices identified in Section 5.1 are provided as examples. The first is an impeller-driven flowmeter/totalizer combination. This type of mechanical device can measure flows from 12 to 380 gpm in a 2-inch pipe.

Accuracy and repeatability are 1% FS and ±0.5% FS, respectively. The power demand typically is 120 VAC or 24 VDC, and the signal output is 4 to 20 mA; both are suitable for DAQ systems.

Drawbacks of this simple type of flowmeter are that it may not be as sensitive as the more expensive devices and it may require more maintenance than a nonmechanical device. Positive features are that it is an insertion-type device, which offers the flexibility of being used for multiple pipe diameters without having the high costs associated with large-diameter meter bodies. Additionally, the meter can be integrated easily with a totalizer. This type of meter would be expected to be easy to install, operate, repair, and replace. An impeller-driven sensor, with flow monitor/totalizer electronics and mounting hardware, would cost about $1,500, placing it at the lower end of the scale for flowmeters.

A second type of flowmeter is the differential-pressure, direct-reading flowmeter. This nonmechanical meter operates on the principle of differential pressure. For 2-inch pipe, flow ranges of 15 to 100 gpm, 20 to 150 gpm, and 30 to 200 gpm are available. The accuracy of the flowmeter reportedly is ±3% FS, which is slightly higher than the accuracy of the impeller, vortex shedder, or magnetic models, but the cost is approximately half that of the magnetic models. External power demand is 24 VDC ±10%, and the electronics component has a signal output of 4 to 20 mA; both are within expected power and RTS DAQ limits.

This flowmeter design carries the cost burden associated with a housing that is the size of the pipe to which it is mounted. However, because it has no moving parts, it reportedly provides extended service with minimal maintenance and repairs. The cost of the flowmeter and its accompanying electronics, which provide an analog output for both rate and total flow, is $1,755.

When selecting a flowmeter/totalizer, the pump output rating, pipe size, sensitivity requirements, installation costs, and durability should be considered.

8.2.2.2 Selecting Pressure Measurement Sensors

For single-point pressure measurements, such as pressure in a delivery pipe either upstream or downstream from a pump, a suitable sensor would be an industrial-grade, general-purpose stainless steel diaphragm semiconductor, as identified in Section 5.2. These sensors are available for pressures from 0 to 500 psig with a proof pressure rating of 200% or 13,000 psi. Accuracy for these sensors is ± 0.25% FS based on linearity, hysteresis, and repeatability. Sensors are temperature-compensated for ranges that should be suitable for both indoor and outdoor facilities. Power demands of AC or DC with millivolt or milliamp outputs are typical, and the sensors are expected to be integrated easily with the RTS power supply and DAQ components. These sensors can be connected to the piping via a ¼-inch NPT fitting which should provide easy removal for cleaning or replacement.

These sensors cost in the $100 to $500 range, with an additional charge of $10 for a snubber to protect the sensor from fluid spikes. An electronic meter designed for on-site readings would cost an additional amount (approximately $325), but it may not be necessary to have one at each facility. If this type of sensor is used at multiple sites, a single meter could be used for real-time readings when an operator visits the site.

For differential-pressure measurements (e.g., for membrane operations), either two single-point sensors or one differential-pressure sensor could be used. Piezoresistive and capacitive differential-pressure sensors are examples of appropriate devices. These sensors have comparable accuracy levels of ±0.5% FS, pressure ranges, power demands of 24 VDC, and outputs of 0 to 5 VDC. The major difference is the acceptable working pressure. One type measures differential pressures at 0 to 200 psid with static operating pressures limited to 300 psi, whereas another unit measures differential pressures from 0 to 100 psid and will tolerate working pressures up to 1,000 psid. Because pre-membrane pressures could range from 50 to 200 psi, with pressure increasing to 100 to 400 psi during fouling, a higher-pressure-rated sensor would be required. Higher pressure ratings generally are reflected in increased costs when selecting a differential-pressure sensor.

8.2.2.3 Selecting Finished Water and Wastewater Production Flowmeters/Totalizers

For RTS applications, electronic totalizers are available for volume measurements, because they will permit remote access and data transmission. The totalizers identified in Section 5.3 are discussed along with flowmeter options in Section 8.2.2.1.

8.2.2.4 Selecting System Status Electronics

As stated in Section 5.4, many electronics manufacturers offer programmable smart relays. These devices provide system status with up to 100 relays, counters, and timers and cost about $275. If electricity is being used by a pump or an electrically driven valve, an addressed relay will be energized to indicate the operation of that particular function. These "on" or "off" flags will make system status monitoring a programmable option for the RTS.

8.2.2.5 Selecting Monitoring Components for Hours of Operation

To monitor the hours of operation, either the programmable relays as discussed in Sections 5.5 and 8.2.2.4 or an independent timer and counter can be used. Timers and counters cost about $40 to $200. The selection of these devices should be based on the voltage delivered to the pump and the output signal available for DAQ and remote telemetry.

8.2.2.6 Selecting Level-Measurement Sensors

Two types of level-measurement devices, identified in Section 5.6, are provided as selection examples. The first sensor is an intrusive RF capacitance sensor, which is a switch that produces a 4-mA output when dry and a 20-mA output when wet. It has an accuracy level of ±1 mm in water, with a power supply of 12 to 36 VDC. It reportedly is impervious to variations in temperature, pressure, light, particles, liquid color, or fluid density. These features for a simple, compact (3-inch-long × ¾-inch-diameter) sensor, which is factory-calibrated with reliability over a wide range of polar liquids, make it an option for RTS use. Installation in a tank is on a 6-foot-long track mount where two sensors would provide "high" and "low" tank signals. If intermediate-level indications are of interest, additional sensors can be installed on the track. The cost for each sensor is $105 with the additional cost of $120 for the commercially available 6-foot-long track mount. Disadvantages of this sensor are that small incremental changes in the level of fluid in a tank are not practical because of the need for many sensors to accomplish this monitoring.

A second example of a level sensor is an ultrasonic sensor that will permit continuous monitoring and will not be in contact with fluid in the tank. The noncontact feature of this sensor reportedly increases the possibility of trouble-free operations. The ultrasonic sensor may be preferred over mechanical float/reed switches or submerged electrode sensors that could become encrusted with dissolved salts which could compromise their performance during extended unattended operations.

One type of ultrasonic sensor can report changes in level from 0.5 to 24.5 feet from the sensor face. The accuracy of the measurements is ±0.25% FS with ±0.125-inch repeatability. Its 24-VDC power demand and 4- to 20-mA signal output are in line with the power demand and output of other sensors and are compatible with anticipated RTS DAQ requirements. The sensor has ATC across its working range of −10 to 140°F, which is compatible with expected operating conditions. Being externally mounted, it can be accessed easily for inspection. No additional electronic meters are expected to be needed because direct interface to the DAQ features of the RTS will provide off-site data. On-site level determination is expected to be visual.

8.2.2.7 Selecting Pumps and Valves

When selecting a pump to incorporate into an RTS, the first consideration is the matrix of the solution to be transferred by the pump. The pump head and any transfer galleries should be inert, yet durable enough to accommodate the liquid being transferred. If the solution is expected to be at either extreme of the pH range, either very acidic or very basic, the pump must be rated to handle material in that pH range. Next, the pump must be able to tolerate the pressure expected to be exerted on the system.

The pump's delivery rate is another consideration. For example, if a buffer needs to be provided to a sample chamber and the volume of this buffer is in the low-milliliter range, a pump with an average delivery rate of liters per minute would not be adequate. A pump of this capacity would have to be run for such a short period of time to deliver a small volume that the metering accuracy probably would be compromised. A pump that provides a low delivery rate could operate at its maximum efficiency over a longer period of time, thus increasing the accuracy of volumetric measurement. How long the pump has to run to provide the correct volume generally is dictated by how quickly the addition needs to be made for a realistic volumetric measurement. Water treatment measurements generally require 1 to 5 minutes to deliver a reagent. If larger volumes need to be transferred, a pump with a higher output rate would be selected, once again to maintain a reasonable transfer time.

The electrical features of COTS pumps, including the power required for operation and any on/off switching features, should be considered. The cost of the various pumps that fit all of the above criteria typically is the final criterion in selecting the COTS pump to be purchased.

The selection of valves follows criteria similar to those for pumps. The valve should (1) be compatible with the solution being transferred, (2) be rated for the expected pressure to be exerted on the valve, (3) have fittings suitable to allow connection of the water delivery lines, (4) contain an orifice that will allow the required flowrate, (5) meet available power supplies, and (6) be cost-effective.

Another feature to be addressed is whether the valve is to be operated in a "normally open/normally closed" function. The "normally open/normally closed" feature indicates the position the valve assumes when not energized. If flow through a valve is to be continuous, except when an instrument or sample needs to be isolated, a "normally open" valve would be recommended. The valve would have to be energized only during short periods of time to terminate flow, which extends the life of the valve and reduces power consumption. If flow is delivered for only short periods of time to some component of the RTS, a "normally closed" valve would be appropriate.

The last feature to be considered is whether the valve should be a piston pilot-operated or piloted-diaphragm valve, or a direct-acting valve. If the valve requires differential pressure to operate, a piston pilot-operated or piloted-diaphragm valve can be used. The piston pilot-operated valve can provide high flows at moderate pressures, whereas the piloted-diaphragm valve offers the highest flows and generally takes up less space. The direct-acting valve requires no differential pressure to open because a coil directly actuates the valve.

8.3 Data Acquisition Systems

8.3.1 Selection Criteria

When selecting DAQ systems for the RTS, the following features should be considered:

- **Amenability/Compatibility.** Will the DAQ component be compatible with the water quality sensors and process monitoring components, and will it collect data that can be accessed remotely with an RTS?
- **Power and I/O Requirements.** Are power demands and I/O requirements suitable to accommodate a wide variety of signal-producing sensors/monitors?
- **Flexibility.** Are the COTS components expandable to allow additional sensors, monitors, and process control options as needed without having to redesign the system? Do the individual components permit upgrading?
- **Software Existence.** Does suitable software exist to interface the hardware to nonvolatile memory that can be retrieved remotely?
- **Remote Diagnostics.** Will the components permit diagnostics for remote identification of repair and maintenance needs?
- **User-Friendly.** Will the integration of the selected hardware and software result in a package that will be user-friendly and be used by operators who may have little experience with computers?
- **Cost.** Will the cost be such that the overall cost of the RTS will be acceptable compared to manual inspection of the facility?

8.3.2 An Example of a Data Acquisition System

A simple, robust, low-cost, and easily maintained operating system should be used for the remote monitoring/control system. The system should provide a suitable real-time format so that its output for a given input occurs rapidly enough for the system to remain within its normal operating parameters. The application should also provide interrupt latency, i.e., the time from the generation of an interrupt until it is acted upon by the operating system, that is suitable for the response/control regime associated with the RTS.

The selected system also must allow easy error recovery. It is expected that errors will occur in the operations at the treatment plants being monitored. Typical errors are blackouts/brownouts, math errors (division by zero), sensor malfunctions (invalid/no response), and hardware failures. An operating system used for control should be able to smoothly recover from these anomalies with minimal impact on the process being controlled.

The operating system is the underlying software that provides a predefined set of tools and services that control system (run time) uses for data input and output, storage, display, and timing. The operating system must be built into the control system, and can be either a simple format such as DOS, or more complex such as Windows® for Intel processors. In general, the more complicated the operating system, the longer/more defined its interrupt latency is and the worse its error-recovery capability becomes.

A DOS-based system provides a robust base for developing a control system. Many low-cost variants are available that contain an extensive set of tools for code development and debugging. DOS generally does not require extensive hardware resources, has documented interrupt latency, and with the proper trapping of errors, is very fault-tolerant. Programming skills required to interface a control program with DOS are minimal. Most programmers who work in the Intel processor family already have these skills. The Windows® environment also has many tools for code development and debugging, but the underlying system at times suffers from ill-defined interrupt latency and poor error recovery. Windows® generally requires significant hardware resources and specialized programming skills to properly interface a control program into the Windows® environment. The newer versions also display some difficulties in trapping error conditions.

The requirements of the control system at a remote site should be different from those at the central monitoring point. The remote site needs a simple, robust, low-cost, and easily maintained control system with the sole purpose of controlling the monitoring/treatment system and providing information to the central system when requested. The remote system should be able to run unattended for long periods of time and restart itself after any interrupt situations, such as a temporary power loss.

Because of these factors, a DOS-based system would be a viable option for RTS applications. The operating system should run from nonvolatile memory (flash or EEPROM) instead of from magnetic media (i.e., disk), because flash/EEPROM has been shown to be highly reliable. The data collected by the remote system will be buffered to a disk, battery backed-up RAM, EEPROM, or flash memory until data transfer is requested by the central computer. Upon request, the buffered data will be transferred, using a serial protocol, to either the central monitoring computer or a laptop on site using simple American Standard

Code for Information Interchange (ASCII) encoding. The central computer or laptop can use the remote site's ASCII-encoded data with a choice of commercial software packages for analysis, plotting, and graphics as required. The central monitoring computer also can be used to make any required adjustments to the remote site control loop.

The central monitoring system can run under whatever operating system is preferred because it will be operated by trained individuals. For reliable RTS operations, one option is to merge a simple DOS-driven software system at the remote site with commercially available data-handling software packages at the central monitoring location.

DAQ boards are available that plug directly into PCs, providing built-in channels to which individual sensors can be attached. However, if one channel develops a problem, the whole board has to be replaced. This would be costly and time consuming. As an alternative, mounting racks with I/O modules may be a better choice. Each module mounts on the board separately. In the event of a single-channel/module failure, the rapid replacement of one new module would correct the problem. COTS mounting racks with I/O modules are readily available and are expected to last for years. Simple mounting boards permit the CPU to be interfaced to a 485 bus. The module format also allows local data processing, which provides the option of control features.

8.4 Remote Telemetry System Options

8.4.1 Selection Criteria

The choice of a suitable telemetry system for an RTS depends on the utilities available at each facility. As the facility becomes more remote, the criteria for telemetry change. For facilities with POTS service, telemetry criteria reflect components utilizing simple modem data transfer. When selecting a telemetry system for this option, the following features should be considered:

- **Compatibility.** Does the system include a modem that is compatible with the DAQ components?
- **Speed.** Does the system have sufficient data transmission speed to permit rapid transfer of data?
- **Environmental Conditions.** Will the system be able to tolerate the environmental conditions at the facility?
- **Availability.** Are system components readily available for easy replacement, if damaged?
- **Cost.** Is the system cost effective?

For more remote facilities where POTS lines do not exist, the additional function of getting the RTS signal from the facility to available phone lines is expanded with the introduction of a cellular phone component. If the facility is even more remote, the above criteria would be applied to RF or satellite options.

8.4.2 Viable Telemetry Options

POTS with modems are an excellent choice if hard-wired transmission lines are available. Modems are relatively inexpensive COTS items that continue to be developed and optimized. For sites where telephone lines are only a few miles away, RF transmission or cellular phone service is possible to relay data to the existing telephone lines. RF transmission is preferred at locations where an RF infrastructure is already in place.

For extremely remote RTS sites, satellite transmission would be the likely option. Data are transmitted by radio waves to satellites which, in turn, relay the signals to a receiving station via telephone lines or a personal dish antenna. Depending on the range of transmission, LEO, MEO, or geostationary satellites are available. The power required to transmit data to geostationary satellites is higher than that for LEO satellites. For a range of about 500 miles, LEO satellites should be adequate. Specific telemetry options will have to be selected on a case-by-case basis for each RTS location.

9

Research Needs

As a result of the literature search, deficiencies in COTS sensors and monitors for RTS applications for small treatment systems have been identified. Only for specific water quality parameters is there a lack of monitoring options. Also, costs of many online monitoring devices are high; many of these devices are retrofitted versions of laboratory-grade instruments. The water quality parameters that may not be easily monitored and controlled include microbial contaminants, TOC, iron and manganese, and color. The research needs for these parameters are discussed in Sections 9.1 through 9.4.

9.1 Microbial Monitoring

A number of analytical techniques for the detection and viability assessment of pathogens and surrogate indicators currently are under development. Most of these methods are based on the use of various types of molecular techniques including immunodetection, nucleic acid-based techniques, and the amplification of nucleic acid by polymerase chain reaction (PCR). Most of these techniques have not been assessed to address the interference by various water quality parameters, such as suspended solids, total and dissolved organic matter, and pH. An attractive possibility is to include a sample separation step based on the use of MF or UF to accomplish separation of microorganisms from potentially interfering substances such as dissolved organic matter.

9.1.1 Immunodetection Methods

Biological detection systems are being developed at Battelle's Columbus Laboratories using the principle of immunodetection. These systems detect cell surface antigens (specific proteins) by attaching the antigens to antibodies. An alternative approach is to use antibodies tagged with a fluorescent/colorimetric group. Three specific sensing systems are described in sections 9.1.1.1 through 9.1.1.3.

9.1.1.1 Biorefractometer™ Polarization Interferometer Biosensor

Development of a Biorefractometer™ biosensor capable of performing a rapid, quantitative, sensitive, label-free immunoassay is ongoing (Boiarski, 1998). The device initially was designed to detect biowarfare agents. This breadboard device can perform a simultaneous immunoanalysis for multiple analytes and it employs a manually interchangeable flow cell.

The technology uses an optical technology, i.e., polarization interferometry, to measure the changes of index of refraction on the surface of a planar single-mode waveguide. Measurement specificity is provided by an antibody coating attached to the waveguide sensor's surface. No tag materials are required because this biosensor methodology measures antigen-antibody binding directly. The binding events change the index of refraction on the surface layer, which alters the velocity of light transmission in the waveguide through its evanescent field interaction. This change is monitored directly, as it occurs, using the polarization interferometer.

The polarization interferometer uses two perpendicular polarization modes of laser light traveling in the waveguide to perform a homogeneous immunoassay (see Figure 9-1). The guided light polarized in a direction perpendicular to the surface (TM mode) has a high degree of evanescent wave interaction with the sensor coating. This vertically polarized light forms the active leg of the polarization interferometer. The horizontally polarized light (TE mode) does not interact with the coating, so it forms the interferometer's reference leg. Antigen-antibody binding on the surface causes a phase shift to occur between the two modes, and this phase shift is measured interferometrically after the light beam leaves the waveguide. The degree of phase shift is proportional to that of the index change, which, in turn, is proportional to the degree of antigen-antibody binding that occurs at the surface. Therefore, the term Biorefractometer™ biosensor applies to this refractive index approach to biosensing.

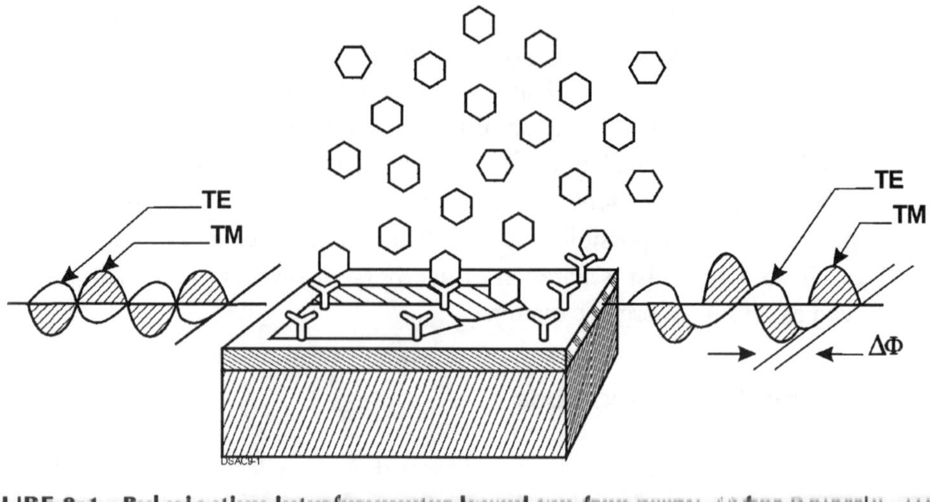

FIGURE 9-1. Polarization Interferometer-based tag-free assay. (After Boiarski, 1998.)

The polarization interferometer can be used to detect multiple analytes. Strips of different antibodies (three strips are shown in Figure 9-2) are placed parallel to each other along the sensor's length. A sample is exposed simultaneously to all three antibody strips. The polarization interferometer's light beam travels under one antibody layer and detects, by phase shift in the light, if any antigen-antibody binding occurs for a particular antibody strip.

Various approaches can be used to sequentially interrogate the antibody strips. One approach would be to move the sensor in a direction perpendicular to the beam to allow sequential interrogation of the antibody strips. The other approach would be to use multiple beams for simultaneous detection of antigen-antibody binding for all antibody strips. Figure 9-2 also shows an opto-electronic (O-E) detector, which can determine phase shift of the beam exiting the waveguide.

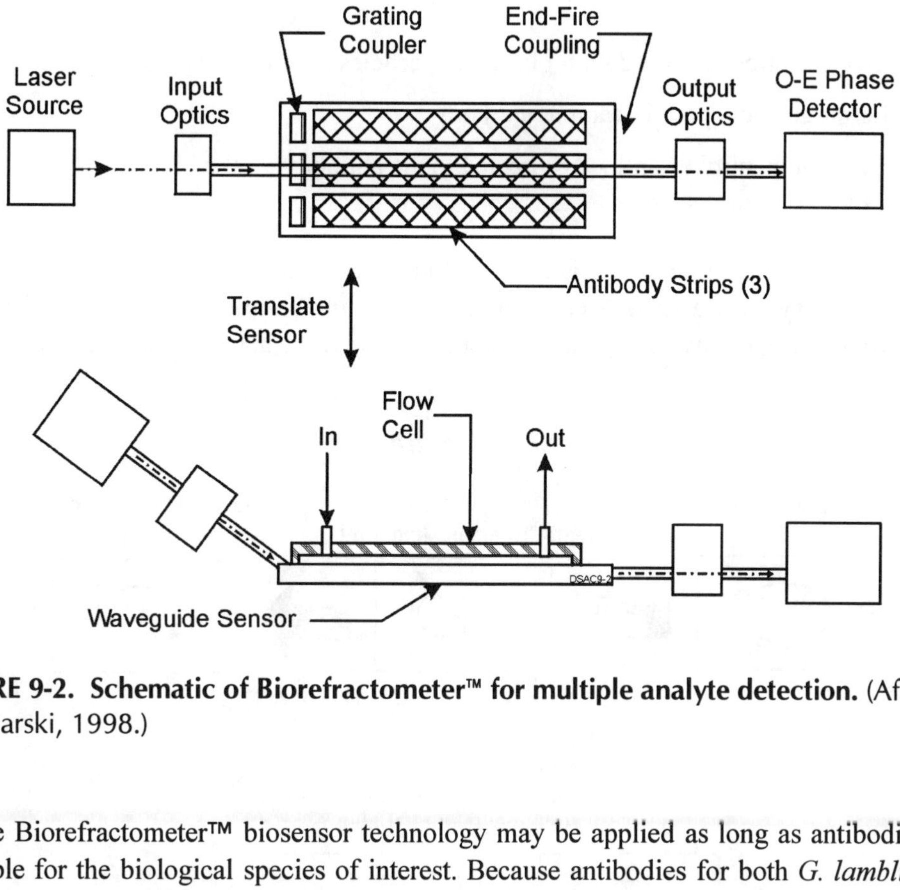

FIGURE 9-2. Schematic of Biorefractometer™ for multiple analyte detection. (After Boiarski, 1998.)

The Biorefractometer™ biosensor technology may be applied as long as antibodies are available for the biological species of interest. Because antibodies for both *G. lamblia* and *C. parvum* are available, it should be possible to monitor these microorganisms. Research into enhancing the sensitivity and further automation of the system is ongoing at Battelle.

9.1.1.2 Flowthrough and One-Step Chromatographic Screening Assays

Another measurement technique that relies on antigen-antibody binding is the handheld screening assay. This technique is being developed by the Naval Medical Research Institute

(NMRI) and Battelle for a wide range of bacterial and toxic agents. Two types of screening assays have been developed. The first, called the flowthrough assay, consists of a sample well and an absorbent pad-backed membrane (see Figure 9-3). When an antigen-containing liquid is added to the sample well, the fluid is absorbed through the membrane and onto the pad and the antigen is captured on the membrane. A reagent solution containing an antibody tagged with an indicating molecule (e.g., colloidal gold) is then added to the sample well and allowed to flow through the membrane surface. When the tagged antibody-antigen complex is captured by the membrane-bound antibody, a color change is produced. When a red "minus sign" appears, it indicates that the test is valid. If an agent is present in the sample, a red "plus sign" will appear. This technique is a qualitative indicator of the presence of the target analyte and would be applicable to *G. lamblia* and *C. parvum* detection.

The second type of screening assay is one-step chromatography. In this case the membrane strip is printed with three lines, as shown in Figure 9-4:

- Detector antibody coated with blue latex particles or colloidal gold
- Microorganism-specific capture antibody
- A reaction control of antibody directed to the antibody on the blue latex particles or colloidal gold.

This triple-coated membrane strip is dried and mounted into a holder called a "ticket." To perform the assay, only a single step is required. After the sample is added to the device, it wicks through the strip, hydrating the detector antibody. If the antigen is present, an antigen-

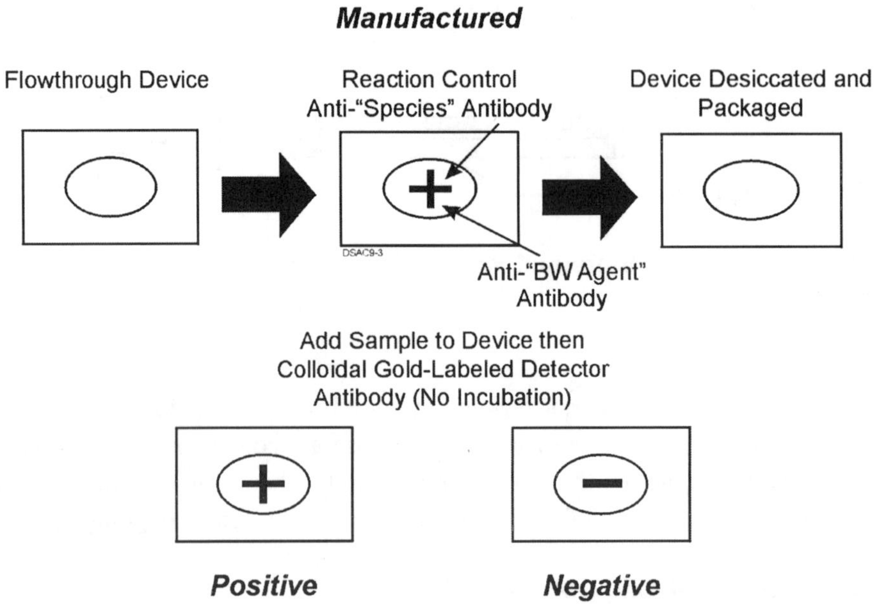

FIGURE 9-3. Configuration of a flowthrough assay and readout of a flowthrough device. (After Boiarski, 1998.)

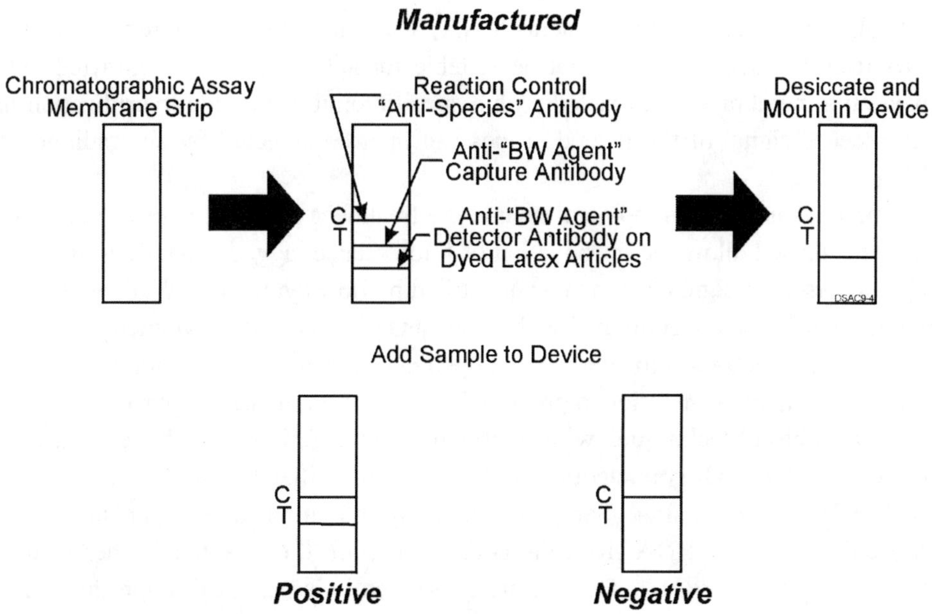

FIGURE 9-4. Readout of a chromatographic assay device. (After Boiarski, 1998.)

antibody complex forms that will be captured by the microorganism-specific capture antibody to form a line in a test window (T). A line also will appear at the site of the reaction control window (C). A negative assay will have only a single line at the reaction control window.

The one-step chromatographic technique initially was designed as a qualitative test. Currently, Battelle is developing a color reader to expand its capability for quantitative measurements. With further development, this technique has the potential for being used to monitor *G. lamblia* and *C. parvum* through an RTS.

9.1.1.3 Magnetic Tagging

The magnetic tagging technology also relies on the antibody affinity technology. An organism-specific antibody is labeled with a magnetic tag that binds to a specific organism and then can be extracted, permitting organism-specific quantitation. It has been indicated (Mariñas, personal communication, 1998) that this technology is similar to other ongoing research into molecular probes and, even though in principle it is feasible, several years of research and development are still needed before it will be possible to concentrate samples sufficiently to permit low-level detection and quantitation of microorganisms such as *C. parvum* using this technique.

9.1.2 Nucleic Acid-Based Techniques

Nucleic acid-based techniques are based on the identification of signature sequences present in the genes encoded in the deoxyribonucleic acid (DNA) of target organisms (Amann et al., 1995). The most common signature sequence for microbial identification is the small subunit (SSU) ribosomal ribonucleic acid (rRNA). Current methods require the extraction of the rRNA from the cell and might not be suitable for automation. The extracted rRNA is then fixed on a membrane and reacted with a specific oligonucleotide hybridization probe. The fluorescence signal of the tagged probes can then be detected by autoradiography or phosphor imaging.

Other methods under development include the treatment of samples to increase the permeability of cells and allow the introduction of fluorescence-tagged probes into microorganisms. The fluorescence signal can then be quantified by flow cytometry (Wallner et al., 1997) or fluorescence microscopy combined with image analysis. The flow cytometry method may require sample concentration to achieve pathogen densities above the practical quantitation limit of the instrument, as mentioned previously for the immunoassay approach.

Nucleic acid-based techniques with microchip configurations also have been reported (Guschin et al., 1997). Oligonucleotide probes are immobilized in a matrix of polyacrylamide gel pads on a glass slide. Samples containing the target microorganism are treated to release either DNA or RNA from the cells, which are then treated further to develop fluorescent-labeled SSU rRNA. The sample subsequently is passed through the microchip where hybridization takes place. The fluorescence signal can then be detected.

The nucleic acid-based techniques present some potential for automation. However, interferences by water quality parameters in the various steps of the techniques remain to be assessed.

9.1.3 Polymerase Chain Reaction

Use of PCR in conjunction with nucleic acid probes presents an attractive technique for the detection and viability assessment of pathogens (Delabre et al., 1997). Current PCR methods require the extraction of nucleic acid (DNA or RNA) from cells, treatment by the various steps described subsequently, and separation by gel electrophoresis. The extracted DNA double strands are first denatured at 90°C to produce single strands. The resulting sample is then annealed to attach conjugate oligonucleotide primers, and subsequently extended with polymerase. The result is that a second strand is formed next to each original single strand, thus doubling the DNA concentration. This process is repeated many times until reaching a desirable DNA concentration. The DNA strands are then separated and identified by gel electrophoresis. This last step currently is the main impediment to the automation of the PCR method. Alternative approaches based on the attachment of fluorescence-tagged probes are under development.

The potential of the PCR technique to assess microbial viability also is under investigation. In general, relatively higher concentrations of the more unstable RNA are viewed as indicators of cell viability.

9.1.4 Biological Surrogate Indicators

The presence of certain pathogens in natural water tends to be associated with fecal contamination of human or animal origin. Because molecular techniques are specific to each target pathogen, simultaneous monitoring of a large number of microorganisms might not be realistic. An alternative approach could be the use of a surrogate indicator, e.g., *Escherichia coli*, to detect fecal contamination. Once automated detection methods, such as the molecular techniques described above, become available, only the indicator microorganism needs to be detected. One disadvantage of this approach is that certain pathogens might be present while the indicator microorganism is absent or present at concentrations below the detection limit of the analytical technique.

9.1.5 Nonbiological Surrogate Indicators

Another innovative approach with great potential for automation and remote sensing is the use of nonbiological surrogate indicators such as fluorescent-dyed polystyrene microspheres. The removal of these indicators by various separation processes may be calibrated with those for various pathogens. The size of the microspheres may be selected to attain optimum correlation with each target microorganism. More importantly, the microspheres can be used to assess the inactivation potential of various disinfection processes (Chiou et al., 1997; Mariñas et al., 1997). The rate decay in fluorescence resulting from the reaction between ozone and the fluorescent chemical inside the microsphere has been correlated successfully with the inactivation kinetics of pathogens such as *Giardia* and *Cryptosporidium* species. Preliminary results from ongoing research have revealed that the fluorescence decay exhibited by the nonbiological surrogate indicators exposed to free chlorine also correlates well with the inactivation kinetics of *Giardia* and *Cryptosporidium* species. The number of microspheres and the fluorescence distribution after separation and disinfection can be assessed by flow cytometry.

9.2 Total Organic Carbon Monitoring

The TOC instruments identified during the search are essentially retrofitted analytical laboratory-grade systems. They have high power demands for their catalytic oxidation reactors, can utilize considerable volumes of consumables, and are high-cost instruments. Two online technologies are being developed that may have potential for RTS applications in the future. One uses optical and fiber optic sensors (Section 9.2.1); the other uses microchip sensors (Section 9.2.2).

9.2.1 Optical and Fiber Optic Sensors

A technology being developed for aqueous TOC monitoring is through optical and fiber optic sensors. Sensors consist of a short length of fiber optics and a light-emitting diode (LED). The fiber is clad with a coating that has an affinity for hydrocarbons. When the fiber is immersed in water, light from the LED passes through the fiber and is detected by a photodiode at the opposite end. In the presence of hydrocarbons, molecules are attracted and attached to the cladding, causing a change in the refractive index of the coating (so that it more closely matches the index of the fiber core). Light emitted from the LED can now escape through the coating, reducing the amount of light detected by the photodiode. The reduction of transmitted light is directly proportional to the hydrocarbon concentration in the surrounding medium. Generally, this type of fiber optic system cannot distinguish between individual hydrocarbons and must be calibrated relative to a single compound. This fiber optic method currently is not suited for online monitoring at a water treatment facility because of limitations in sensitivity, the need to clean and restore the integrity of the cladding, and signal drifts.

9.2.2 Microchip Sensors

Microchip sensors are being developed primarily for gaseous hydrocarbon monitoring, but they may be developed to also address the aqueous matrix. Research is being done at the National Aeronautics and Space Administration's Jet Propulsion Laboratory in Pasadena, California, where ions produced during the catalysis of hydrocarbons associate themselves on a silicon-carbide layered chip to permit both qualitative and quantitative analyses (*Electronic Design*, 1997). Although this technology reportedly is in the demonstration-of-principle stage, it may produce a sensor option in the future.

At Los Alamos National Laboratory, research is ongoing to build a selective chemical sensor for chlorinated hydrocarbons. This microchip sensor uses surface-acoustic-wave techniques on a thin silicon-nitride membrane. The chip contains a binding chemical that causes the hydrocarbon of interest to stick to the membrane, thus altering the surface acoustic signature and permitting measurements of concentrations for that specific compound.

9.3 Metals Monitoring

The primary market for metals analyses employs laboratory instruments, such as atomic absorption (AA), inductively coupled plasma (ICP), and ion chromatography (IC). Therefore, developmental efforts made by most manufacturers are focused mainly on refining these instruments rather than developing new, compact, and inexpensive online monitors. The only area of research identified for metal monitoring in water is the optical route. Fiber optic sensors reportedly are being investigated for possible online monitoring of not only metals, but nitrates and chlorine, which are all of interest for water quality monitoring (Bogue, 1994). The optical researchers imply that it may be possible to perform metals analyses with fiber optic technology, but they provide no indication that any COTS devices will be available soon. The need for developing inexpensive metal sensors apparently still exists but is not being market-driven due to low demand.

9.4 Color Monitoring

Limited efforts to advance color monitoring have taken place. A discussion on the development of a comprehensive water quality monitoring system for Space Station Freedom (Niu et al., 1990) indicates that the recommended technology to measure color would still be spectrophotometry, even under the constraints of power, weight, and instrument size associated with space vehicles. The only newer technology being investigated is fiber optics, which may offer an option for color monitoring in the future. No definitive research, however, was identified as a result of the literature search.

9.5 References

Amann, R.I.; Ludwig, W.; Schleifer, K.H. (1995), Phylogenetic Identification and In Situ Detection of Individual Microbial Cells without Cultivation, *Applied and Environmental Microbiology, 59(1),* pp. 143-169.

Bogue, R. (1994), Sensor Advances for Water Quality, *Control and Instrumentation, 26(6),* p. 36.

Boiarski, A.A. (1998), *Survey of Existing Biosensor Instrumentation*, Battelle Memorial Institute, Columbus, Ohio.

Chiou, C.-F.; Torres-Lugo, M.; Mariñas, B.J.; Adams, J.Q. (1997), Non-biological Surrogate Indicators for Assessing Ozone Disinfection, *Journal of the American Water Works Association, 89(8),* pp. 54-66.

Delabre, K.; Mennecart, V.; Joret, J.-C.; Cervantes, P. (1997), Simultaneous Detection of Pathogenic Bacteria in Water Using Nested-PCR, *Proceedings of the American Water Works Association Water Quality Technology Conference,* November 9-12, 1997, Denver, Colorado.

Electronic Design (1997), Vol. 45, Number 14, p. 35, Penton Publishing, Inc., Cleveland, Ohio.

Guschin, D.Y.; Mobarry, B.K.; Proudnikov, D.; Stahl, D.A.; Rittmann, B.E.; Mirzabekov, A.D. (1997), Oligonucleotide Microchips as Genosensors for Determinative and Environmental Studies in Microbiology, *Applied and Environmental Microbiology, 63(6),* pp. 2397-2402.

Mariñas, B.J. (1998), Personal communication with B.J. Mariñas, University of Illinois, Urbana-Champaign, Illinois. March.

Mariñas, B.J.; Rennecker, J.L.; Hung, C.-H.; Teefy, S.; Rice, E.W.; Owens, J.H. (1997), Assessment of Ozone Disinfection Efficiency Using Fluorescent-Dyed Polystyrene Microspheres as Indicators for *Cryptosporidium* Oocysts, *Proceedings of the 1997 American Water Works Association Annual Conference, Volume C: Water Research*, June 15-19, 1997, Atlanta, Georgia, pp. 531-544.

Niu, W.; Burchfield, D.; Snyder, G.; Conklin, K. (1990), Development of a Water Quality Monitor for Space Station Freedom Life Support System, *Society of Automotive Engineers Transactions, 99,* pp. 1256-1265.

Wallner, G.; Fuchs, B.; Spring, S.; Beisker, W.; Amann, R. (1997), Flow Sorting of Microorganisms for Molecular Analysis, *Applied and Environmental Microbiology, 63(11),* pp. 4223-4231.

A

Literature Search Strategy

The open literature was reviewed to obtain information about vendors of water quality monitoring devices and services for small communities, supervisory control and data acquisition (SCADA) equipment, remote telemetry applications used by electric power and natural gas utilities, and RTS packages. The literature review was conducted using the DIALOG OneSearch system. The Internet was searched for additional information.

Appropriate technologies were identified using the search, and telephone interviews were conducted with the manufacturers and vendors. The water quality parameters included turbidity, pH, chemical dose/residual concentration, temperature, TOC, trace metals, anions/cations, color, hardness, TDS, and particle counts. The system operating parameters included flowrate, pressure, electric usage, amount of drinking water produced, amount of wastewater produced, operating time, and chemical feed levels. The literature search for data acquisition/remote telemetry options included systems being used by the electric power and natural gas utility industries.

Three consecutive searches were conducted. The search strategies for the three searches are discussed in Sections A.1 through A.3.

A.1 Search 1 (Water Quality)

Search 1 concentrated on water quality parameters. The search concentrated on the following scientific databases:

- Engineering Index Compendex
- Information Access Company (IAC) Trade and Industry Database
- U.S. Patents 1980-97

- Federal Register 1988-97
- European Patents
- Toxline 1965-97
- Medline 1966-97
- Energy Science & Technology 1974-97
- National Technical Information Service (NTIS)
- IAC Newsearch
- Commercial Agriculture Bureau (CAB) Abstracts.

Additional journals were searched:

- Textline Global News
- Federal News Service
- Business Dateline
- Food Science and Technology Abstracts
- American Medical Association (AMA) Journals
- Biosis Previews
- Analytical Abstracts
- New England Journal of Medicine
- Information Handling Systems (IHS) International Standards and Specifications
- Wilson General Science.

The following keywords were used for the search:

- Drinking/potable water (I)
- Sensor/monitor (II)
- The water quality parameters listed above (III)
- Software to control, DAQ/log, data transmission, and remote telemetry (IV).

The search was done by targeting articles that contained keywords in I, II, and any one of the water quality parameters listed in III. This was followed by a search for articles that contained keywords in I, II, and any of the keywords in IV.

A separate search was carried out using the following trade magazine databases:

- IAC Trade & Industry Database 1976-97
- ABI/INFORM 1971-97
- IAC PROMT 1972-97
- IAC MARS 1984-97
- IAC Globalbase
- FINDEX.

To further identify pertinent information, the following specific trade magazines were searched:

- Chemical Engineering
- Pollution Equipment News
- Journal of Commerce
- Business and Industry
- Financial Times
- Freedonia Market Research
- Industry Trends and Analysis
- Chemical Industry Notes
- Euromonitor Market Research.

The final major source included in Search 1 was the Thomas Register. It was used to obtain a list of manufacturers/suppliers for sensors/controls used in measuring the target water quality parameters.

A.2 Search 2 (System Operations)

In Search 2, components to track system operating parameters were investigated. Because information was expected to be readily available for these traditional process functions, only the trade magazine database search was conducted. The trade magazine databases, specific trade magazines, and the Thomas Register used in Search 1 also were used for Search 2.

A.3 Search 3 (Data Acquisition and Remote Telemetry)

Search 3 concentrated on both DAQ and remote telemetry systems. The search was focused on the previously identified scientific databases and on the trade magazine databases on DAQ and remote telemetry. The following keywords were used during Search 3:

- Remote telemetry
- Data acquisition
- Data transmission
- Software to control, acquire, transmit data, communicate
- SCADA
- Remote control and operation.

The literature search identified articles that had matches for any of the above six sets of keywords. Articles that also contained the keywords "energy utilities" were located to determine remote telemetry and DAQ methods used by energy utilities. To find articles describing specific remote telemetry methods, the articles that had matches with any of the above five sets of keywords were further defined by using the following keywords:

- Radiofrequency
- Land-based phone system
- Cellular phone system
- Satellite communications.

In addition to this literature search, vendors were manually identified in the trade magazines Sensors, Pollution Equipment News, and Chemical Engineering. This search was used to identify additional vendors for sensors, DAQ systems, and remote telemetry components.

B

COTS Devices for Water Quality Monitoring and Control

The COTS devices for water quality monitoring and control include the various sensors and monitors referenced in the tables presented in Chapter 4.

This listing is provided solely for informational purposes. Inclusion in this appendix does not constitute recommendation of the companies listed or endorsement for use of their products. Reasonable efforts have been made to ensure that the information provided is up to date.

Rosemount Model T/T1054-A2120-FT-M

1) Analyte	Turbidity
2) Method	Light scattered at 90° detected by photodiode
3) Range/Sensitivity	0.000 to 200 NTUs; resolution is 0.001 NTU below 1 NTU
4) Accuracy	±2% of reading from 0 to 60 NTUs and ±5% of reading from 60 to 200 NTUs
5) Repeatability/Life Expectancy	Better than ±1% of reading or ±0.003 NTU or ±1 light-scattering division (LSD), whichever is greater
6) Zero Drift @ Constant Temp.	Not reported (NR)
7) Zero Drift @ Ambient Temp.	NR
8) Zero Noise @ Constant Temp.	NR
9) External Power Supply	12/240 VAC, 50/60 Hz, 11 watts maximum
10) Power Consumption	NR
11) Output Signal	Isolated 0/4 to 20 mA or 0 to 5 VDC with programmable span in any portion of the range in 0.001 NTU or larger increments; optional 0 to 5 VDC also with programmable span
12) Electronic Response Time	Instantaneous initial response for a full step change, 90% response in 30 sec, at 25 gallons per hour (gph) flowrate
13) Operating Temperature Range	0 to 50°C (32 to 122°F); sample temperature range of 0 to 60°C (32 to 140°F)
14) Storage Temperature Range/ Requirements	NR
15) Weight	6.36 kg (14 lb) shipping weight
16) External Dimensions	Flowthrough chamber 4.38" W x 8.25" H x 4.38" D; 0.5" NPTM inlet and outlet connections
17) Calibration Requirements/ Maintenance	Calibrated with Formazin or AEPA-1 polymer standards; also a secondary standard for quick calibration checks
18) Manufacturer	Rosemount
19) Cost	$1,776 for A2100 sensor; $823 for T1054 analyzer
20) Comments	Includes two alarm relays; available with flowthrough or debubbling chamber; integral preamplifier; signal can be carried up to 50 ft to analyzer; photodiode and light source are included in same sensor body; 10 to 100 gph sample flow required

Hach Model 1720D

1)	**Analyte**	Turbidity
2)	**Method**	Light scattered at 90° detected by submerged photocell
3)	**Range/Sensitivity**	0 to 100 NTUs; resolution is 0.001 NTU
4)	**Accuracy**	±2% of reading, or ±0.020 NTU, 0 to 30 NTUs; ±5% of reading, 30 to 100 NTUs
5)	**Repeatability/Life Expectancy**	±1.0% or ±0.002 NTU, whichever is greater
6)	**Zero Drift @ Constant Temp.**	NR
7)	**Zero Drift @ Ambient Temp.**	NR
8)	**Zero Noise @ Constant Temp.**	NR
9)	**External Power Supply**	95 to 240 VAC, 50/60 Hz, autoselect, 40 VA
10)	**Power Consumption**	NR
11)	**Output Signal**	Selectable for 0 to 20 or 4 to 20 mA
12)	**Electronic Response Time**	For a full-scale step change, initial response in 1 min, 15 sec; varies with flowrate.
13)	**Operating Temperature Range**	0 to 40°C (sample temperature of 0 to 50°C)
14)	**Storage Temperature Range/ Requirements**	NR
15)	**Weight**	25 lb
16)	**External Dimensions**	9 x 9 x 7" for AquaTrend interface; 9 x 9 x 7" for power supply; 10 x 12 x 16" for turbidimeter body and cap
17)	**Calibration Requirements/ Maintenance**	Requires minimal routine maintenance; calibration based on Formazin standards
18)	**Manufacturer**	Hach
19)	**Cost**	$950 for turbidimeter with PS1201 power supply; $1,795 for turbidimeter with AquaTrend interface and power supply
20)	**Comments**	Applications include monitoring drinking water down to 0.001 NTU; sample flow of 250 to 750 mL/min required; two high/low turbidity setpoint alarms; meets U.S. EPA established minimum instrument design criteria for turbidity measurements and is suitable for regulatory reporting

Honeywell

1)	**Analyte**	Turbidity, conductivity, temperature
2)	**Method**	Turbidity is measured using a 935 nm infrared LED and two detectors (one transmissive and one scattered)
3)	**Range/Sensitivity**	0 to 4000 NTUs for turbidity; 68 to 140°F for temperature
4)	**Accuracy**	±4°F; see manufacturer's curve for turbidity and conductivity
5)	**Repeatability/Life Expectancy**	NR
6)	**Zero Drift @ Constant Temp.**	NR
7)	**Zero Drift @ Ambient Temp.**	NR
8)	**Zero Noise @ Constant Temp.**	NR
9)	**External Power Supply**	8 to 30 VDC power supply
10)	**Power Consumption**	NR
11)	**Output Signal**	Data transmitted via a bidirectional 5 VDC RS-232 serial communications link
12)	**Electronic Response Time**	1.3 sec for turbidity; 0.03-sec response time and 3- to 5-min stabilization time for temperature; 0.85 sec for conductivity
13)	**Operating Temperature Range**	68 to 140°F; may be wider range; however, has been tested only within this range
14)	**Storage Temperature Range/ Requirements**	NR
15)	**Weight**	NR
16)	**External Dimensions**	Mounting dimension is 1.25" diameter x 2.39"; cap is 0.71" outside diameter (OD) x 2.75"
17)	**Calibration Requirements/ Maintenance**	NR
18)	**Manufacturer**	Honeywell
19)	**Cost**	$90 without flowthrough cap; $100 with flowthrough cap
20)	**Comments**	With microprocessor and software

GLI International 95T/8220

1)	**Analyte**	Turbidity
2)	**Method**	Ratiometric four-beam, U.S. EPA approved GLI Method 2
3)	**Range/Sensitivity**	0.000 to 1.000, 0.00 to 10.00, and 0.0 to 100.0 NTUs with autoranging; sensitivity is 0.001 NTU
4)	**Accuracy**	±2% of reading, all ranges
5)	**Repeatability/Life Expectancy**	0.1% of span or better
6)	**Zero Drift @ Constant Temp.**	0.01% of span per °C
7)	**Zero Drift @ Ambient Temp.**	NR
8)	**Zero Noise @ Constant Temp.**	NR
9)	**External Power Supply**	98 to 132 VAC, 50 or 60 Hz (switch selectable), 15 VA typical; optional 195 to 265 VAC
10)	**Power Consumption**	NR
11)	**Output Signal**	0 to 1 mA (100 ohms maximum load), 0 to 5 VDC (5000 ohms minimum load), and 4 to 20 mA (600 ohms maximum load)
12)	**Electronic Response Time**	1, 6, 30, 60 or 90 sec to 90% of value upon step change, switch selectable
13)	**Operating Temperature Range**	Sensor is 32 to 140°F (0 to 60°C)
14)	**Storage Temperature Range/ Requirements**	Sensor is 32 to 140°F (0 to 60°C); analyzer is −4 to 140°F (−20 to 60°C)
15)	**Weight**	Analyzer, 12 lb (5.5 kg); sensor, 18 lb (8.2 kg)
16)	**External Dimensions**	12.5" W x 9.7" H x 6.6" D
17)	**Calibration Requirements/ Maintenance**	Practical lifetime calibration (only periodic verification of calibration using the optional patented turbid glass cube assembly is required); only required maintenance is occasional cleaning of flow chamber
18)	**Manufacturer**	GLI International, Inc.
19)	**Cost**	$950 for 95T analyzer; $1,050 for 8220 sensor
20)	**Comments**	Built-in diagnostic messages; LED light source lasts 10 to 12 yr; flowrate ranges from 0.05 to 7 gpm; residence time is 9.5 sec at 1 gpm; designed for lower turbidity

GLI International 95T/8224T

1) Analyte	Turbidity
2) Method	Ratiometric four-beam, U.S. EPA approved GLI Method 2
3) Range/Sensitivity	0.00 to 10.00, 0.0 to 100.0, and 0 to 1,000 NTUs with auto-ranging; sensitivity is 0.01 NTU
4) Accuracy	±5% of range
5) Repeatability/Life Expectancy	0.1% of span or better
6) Zero Drift @ Constant Temp.	0.01% of span per °C
7) Zero Drift @ Ambient Temp.	NR
8) Zero Noise @ Constant Temp.	NR
9) External Power Supply	98 to 132 VAC, 50 or 60 Hz (switch selectable), 15 VA typical; optional 195 to 265 VAC
10) Power Consumption	NR
11) Output Signal	0 to 1 mA (100 ohms maximum load), 0 to 5 VDC (5000 ohms minimum load), and 4 to 20 mA (600 ohms maximum load)
12) Electronic Response Time	For a step change, time required to reach 99% of full value is 5 sec in fastest response and 450 sec in slowest response (1-, 6-, 30-, 60-, or 90-sec settings, switch selectable)
13) Operating Temperature Range	Sensor is 32 to 140°F (0 to 60°C)
14) Storage Temperature Range/ Requirements	Sensor is 32 to 140°F (0 to 60°C); analyzer is −4 to 140°F (−20 to 60°C)
15) Weight	Analyzer, 12 lb (5.5 kg); sensor, 24 lb (10.9 kg)
16) External Dimensions	12.5" W x 9.7" H x 6.6" D
17) Calibration Requirements/ Maintenance	Practical lifetime calibration (only periodic verification of calibration using the optional patented turbid glass cube assembly is required); falling stream technology eliminates sensor fouling, providing years of maintenance-free operation
18) Manufacturer	GLI International, Inc.
19) Cost	$950 for 95T analyzer; $2,200 for 8224T sensor
20) Comments	Built-in diagnostic messages; LED light source lasts 10 to 12 yr; inlet flowrate 5 to 9 gpm; residence time 10 sec at 6 gpm; designed for higher turbidity

Global Water WQ700 with EZ100 Display

1)	**Analyte**	Turbidity
2)	**Method**	Incandescent light and 90° light scattering
3)	**Range/Sensitivity**	0 to 200 NTUs; optional 0 to 2,000 NTUs or 0 to 10 NTUs
4)	**Accuracy**	±0.1 NTU for 0 to 200 NTUs
5)	**Repeatability/Life Expectancy**	±0.05 NTU for 0 to 200 NTUs
6)	**Zero Drift @ Constant Temp.**	NR
7)	**Zero Drift @ Ambient Temp.**	NR
8)	**Zero Noise @ Constant Temp.**	NR
9)	**External Power Supply**	12 VDC, 50 mA typical
10)	**Power Consumption**	NR
11)	**Output Signal**	4 to 20 mA (three-wire)
12)	**Electronic Response Time**	NR
13)	**Operating Temperature Range**	−10 to 70°C
14)	**Storage Temperature Range/ Requirements**	NR
15)	**Weight**	2 lb
16)	**External Dimensions**	Body is 1.1" x 6"; shield is 1.7 x 4"; overall length is 8"
17)	**Calibration Requirements/ Maintenance**	Incandescent lamp has 20-yr continuous usage
18)	**Manufacturer**	Global Water
19)	**Cost**	$800 for sensor; $285 for LCD display; $185 for flowthrough chamber
20)	**Comments**	25' standard cable; rugged housing

OMEGA TRCN95

1)	**Analyte**	Turbidity
2)	**Method**	Four-beam ratiometric measurement
3)	**Range/Sensitivity**	0.001 to 100 NTUs; sensitivity is 0.001 NTU
4)	**Accuracy**	±2% of reading
5)	**Repeatability/Life Expectancy**	±0.1% of span or better
6)	**Zero Drift @ Constant Temp.**	NR
7)	**Zero Drift @ Ambient Temp.**	NR
8)	**Zero Noise @ Constant Temp.**	NR
9)	**External Power Supply**	110 VAC, 50 or 60 Hz (switch selectable), 15 VA typical, optional 220 VAC
10)	**Power Consumption**	NR
11)	**Output Signal**	Isolated 0 to 1 mA, 100-ohm maximum load; isolated 0 to 5 VDC, 5000-ohm minimum load; isolated 4 to 20 mA, 600-ohm maximum load
12)	**Electronic Response Time**	1, 6, 30, 60, or 90 sec
13)	**Operating Temperature Range**	Sample temperature range of 0 to 60°C (32 to 140°F)
14)	**Storage Temperature Range/ Requirements**	0 to 60°C (32 to 140°F)
15)	**Weight**	Sensor is 18 lb (approximate); analyzer is 12 lb (approximate)
16)	**External Dimensions**	Sensor is 17.5" H x 13.8" W x 7.6" D; analyzer is 12.5" H x 9.7" W x 6.6" D
17)	**Calibration Requirements/ Maintenance**	Minimal maintenance; occasional cleaning of sample chamber baffle and glass window; calibration with Formazin or optical calibration cube
18)	**Manufacturer**	OMEGA
19)	**Cost**	$1,995; $395 for turbidity calibration cube
20)	**Comments**	Automatic color compensation; bubble-rejecting sample chamber; autoranging; flowrate of 0.05 to 7 gpm; connect to 1/2" NPT female standard (adaptable to 3/8" or 1/4"); surface or pipe-mount installation; setpoints adjustable from 0 to 100% of range; warning and failure diagnostic signals

Environmental Sensors Model 2600

1)	**Analyte**	Turbidity
2)	**Method**	Backscatter nephelometer
3)	**Range/Sensitivity**	500, 1000, 2000 FTUs
4)	**Accuracy**	2% of full scale
5)	**Repeatability/Life Expectancy**	NR
6)	**Zero Drift @ Constant Temp.**	NR
7)	**Zero Drift @ Ambient Temp.**	NR
8)	**Zero Noise @ Constant Temp.**	NR
9)	**External Power Supply**	Voltage 9 to 15 VDC unregulated; current 20 mA plus current loop
10)	**Power Consumption**	NR
11)	**Output Signal**	4 to 20 mA/three-wire
12)	**Electronic Response Time**	1 sec
13)	**Operating Temperature Range**	0 to 30°C temperature compensated
14)	**Storage Temperature Range/ Requirements**	NR
15)	**Weight**	NR
16)	**External Dimensions**	200 mm L x 32 mm diameter (7.875" x 1.25")
17)	**Calibration Requirements/ Maintenance**	NR
18)	**Manufacturer**	Environmental Sensors
19)	**Cost**	$1,200
20)	**Comments**	

GLI International Liquid Crystal Polymer-Encapsulated pH Sensor (6028PO) with Model 672P Analyzer

1)	**Analyte**	pH
2)	**Method**	Differential electrode (process, reference, and ground electrode)
3)	**Range/Sensitivity**	0 to 14 pH; sensitivity less than 0.005 pH for sensor; −500 to 500 mV and −10.0 to 100.0°C
4)	**Accuracy**	Nonlinearity is 0.05% of span
5)	**Repeatability/Life Expectancy**	Stability is 0.03 pH per 24 hr, noncumulative; analyzer repeatability is 0.05% of span or better
6)	**Zero Drift @ Constant Temp.**	0.01% of span per °C
7)	**Zero Drift @ Ambient Temp.**	NR
8)	**Zero Noise @ Constant Temp.**	NR
9)	**External Power Supply**	98 to 132 VAC, 50/60 Hz
10)	**Power Consumption**	24 VDC with option described below
11)	**Output Signal**	Optional built-in two-wire transmitter if 4 to 20 mA signal is required; output span is 0.96 mA per pH unit; 12 mA at 7.0 pH ±0.88 pH; analyzer output is isolated 4 to 20 mA, 0 to 5 VDC, and 0 to 1 mA
12)	**Electronic Response Time**	NR
13)	**Operating Temperature Range**	−5°C to 60°C (for polyvinyl chloride [PVC])
14)	**Storage Temperature Range/ Requirements**	−22 to 122°F (−30 to 50°C), 0 to 95% relative humidity
15)	**Weight**	Analyzer is 3 lb (1.36 kg)
16)	**External Dimensions**	Sensor is 7.1" L with 1.5" NPT threading; analyzer is 5.57" x 5.67" x 5.46"
17)	**Calibration Requirements/ Maintenance**	Automated one- or two-point calibration
18)	**Manufacturer**	GLI International, Inc.
19)	**Cost**	$545 for sensor; $885 for analyzer
20)	**Comments**	Encapsulated construction protects electronics from moisture and humidity; stable with temperature variations; built-in preamplifier; 10 ft/sec maximum flow; analyzer can display pH, temperature, mV signal from sensor, and 4 to 20 mA output signal; automatic temperature compensation from 0 to 100°C (32 to 212°C)

GLI International Model 6070PO

1)	**Analyte**	pH
2)	**Method**	Differential electrode technique
3)	**Range/Sensitivity**	0 to 14 pH; sensitivity is <0.005 pH
4)	**Accuracy**	NR
5)	**Repeatability/Life Expectancy**	Stability is 0.03 pH unit per 24 hr, noncumulative
6)	**Zero Drift @ Constant Temp.**	NR
7)	**Zero Drift @ Ambient Temp.**	NR
8)	**Zero Noise @ Constant Temp.**	NR
9)	**External Power Supply**	NR
10)	**Power Consumption**	NR
11)	**Output Signal**	0.95 mA per pH unit for two-wire sensor; 12 mA occurs at 7.0 pH ±0.6 pH
12)	**Electronic Response Time**	NR
13)	**Operating Temperature Range**	−5 to 60°C or greater depending on mounting
14)	**Storage Temperature Range/ Requirements**	NR
15)	**Weight**	7.5 lb (3.4 kg) for sensor and mounting assembly
16)	**External Dimensions**	Sensor and mounting assembly are 27" long; mounting fits 2" NPT pipe
17)	**Calibration Requirements/ Maintenance**	NR
18)	**Manufacturer**	GLI International, Inc.
19)	**Cost**	$675 for sensor
20)	**Comments**	Automatic temperature compensation; options include sensor with integral preamplifier/five-wire cable or sensor with integral transmitter with two-wire 4 to 20 mA output; maximum flow of 10 ft/sec

Orion Model 91-07

1)	**Analyte**	pH
2)	**Method**	Triode™ 3-in-1 pH ATC probe
3)	**Range/Sensitivity**	0 to 14 pH
4)	**Accuracy**	±1°C for temperature; 0.05 to 0.1 pH unit
5)	**Repeatability/Life Expectancy**	NR
6)	**Zero Drift @ Constant Temp.**	NR
7)	**Zero Drift @ Ambient Temp.**	NR
8)	**Zero Noise @ Constant Temp.**	NR
9)	**External Power Supply**	NR
10)	**Power Consumption**	NR
11)	**Output Signal**	Millivolt
12)	**Electronic Response Time**	Depends on meter
13)	**Operating Temperature Range**	0 to 90°C
14)	**Storage Temperature Range/ Requirements**	NR
15)	**Weight**	<1 lb (electrode with cable)
16)	**External Dimensions**	110 mm x 12 mm (cap diameter 16 mm)
17)	**Calibration Requirements/ Maintenance**	Little or no maintenance needed
18)	**Manufacturer**	Orion
19)	**Cost**	$134 to $150 (depending on connector)
20)	**Comments**	Epoxy body, nonrefillable, portable, with ATC, Ag/AgCl reference type, Wick junction

Solomat 407pHX

1)	**Analyte**	pH
2)	**Method**	Combination electrode
3)	**Range/Sensitivity**	0 to 14 pH
4)	**Accuracy**	±0.2 pH unit
5)	**Repeatability/Life Expectancy**	Approximately 1 yr under normal conditions
6)	**Zero Drift @ Constant Temp.**	NR
7)	**Zero Drift @ Ambient Temp.**	NR
8)	**Zero Noise @ Constant Temp.**	NR
9)	**External Power Supply**	NR
10)	**Power Consumption**	NR
11)	**Output Signal**	Millivolt
12)	**Electronic Response Time**	Depends on meter
13)	**Operating Temperature Range**	0 to 100°C
14)	**Storage Temperature Range/ Requirements**	NR
15)	**Weight**	A few ounces
16)	**External Dimensions**	105 mm x 12 mm (0.47" x 4.13")
17)	**Calibration Requirements/ Maintenance**	Can be refilled but usually are disposed of
18)	**Manufacturer**	Solomat
19)	**Cost**	$199
20)	**Comments**	Plastic housing, gel electrolyte, simultaneous pH and temperature measurement/compensation, 1-m cable

Cole-Parmer General-Purpose

1)	**Analyte**	pH
2)	**Method**	Single-junction electrode
3)	**Range/Sensitivity**	0 to 13 pH
4)	**Accuracy**	NR
5)	**Repeatability/Life Expectancy**	Approximately 2 to 3 yr
6)	**Zero Drift @ Constant Temp.**	NR
7)	**Zero Drift @ Ambient Temp.**	NR
8)	**Zero Noise @ Constant Temp.**	NR
9)	**External Power Supply**	NR
10)	**Power Consumption**	NR
11)	**Output Signal**	Millivolt
12)	**Electronic Response Time**	Depends on meter
13)	**Operating Temperature Range**	−5 to 100°C (23 to 212°F)
14)	**Storage Temperature Range/ Requirements**	NR
15)	**Weight**	0.3 lb
16)	**External Dimensions**	Approximately 4"
17)	**Calibration Requirements/ Maintenance**	Depends on application
18)	**Manufacturer**	Cole-Parmer
19)	**Cost**	$105 to $150
20)	**Comments**	Industrial, 100 ohm (Ω) or 1 kΩ or thermistor as ATC element, blue glass bulb, single-junction; options include mounting assembly and preamplified electrode, bayonet N connector (BNC)

Cole-Parmer Self-Cleaning Electrode

1)	**Analyte**	pH
2)	**Method**	Double-junction
3)	**Range/Sensitivity**	NR
4)	**Accuracy**	NR
5)	**Repeatability/Life Expectancy**	NR
6)	**Zero Drift @ Constant Temp.**	NR
7)	**Zero Drift @ Ambient Temp.**	NR
8)	**Zero Noise @ Constant Temp.**	NR
9)	**External Power Supply**	NR
10)	**Power Consumption**	NR
11)	**Output Signal**	Millivolt
12)	**Electronic Response Time**	Depends on meter
13)	**Operating Temperature Range**	CPVC, 80°C; Kynar™, 100°C
14)	**Storage Temperature Range/ Requirements**	NR
15)	**Weight**	0.45 lb
16)	**External Dimensions**	Approximately 4" in length
17)	**Calibration Requirements/ Maintenance**	Depends on application
18)	**Manufacturer**	Cole-Parmer
19)	**Cost**	$100 to $141
20)	**Comments**	Industrial, self-cleaning, flat electrode, 100 Ω or 1 kΩ or thermistor as ATC element, chlorinated polyvinyl chloride (CVPC) or Kynar™ housing, double-junction; optional tee fitting, BNC connector

Cole-Parmer In-Line

1)	**Analyte**	pH
2)	**Method**	Combination electrode
3)	**Range/Sensitivity**	0 to 13 pH
4)	**Accuracy**	NR
5)	**Repeatability/Life Expectancy**	NR
6)	**Zero Drift @ Constant Temp.**	NR
7)	**Zero Drift @ Ambient Temp.**	NR
8)	**Zero Noise @ Constant Temp.**	NR
9)	**External Power Supply**	NR
10)	**Power Consumption**	NR
11)	**Output Signal**	Millivolt
12)	**Electronic Response Time**	Depends on meter
13)	**Operating Temperature Range**	NR
14)	**Storage Temperature Range/ Requirements**	NR
15)	**Weight**	NR
16)	**External Dimensions**	4" L x 15/16 "-diameter housing; electrode extends 1" beyond
17)	**Calibration Requirements/ Maintenance**	Depends on application
18)	**Manufacturer**	Cole-Parmer
19)	**Cost**	$103 to $124
20)	**Comments**	Disposable, industrial, rugged design, blue glass sensor, Kynar™ housing, single- and double-junction available, optional probe guard, submersible or in-line available, BNC connectors

OMEGA PHE/ORE-5460

1)	**Analyte**	pH
2)	**Method**	Combination electrode
3)	**Range/Sensitivity**	0 to 14 pH
4)	**Accuracy**	0.01
5)	**Repeatability/Life Expectancy**	NR
6)	**Zero Drift @ Constant Temp.**	NR
7)	**Zero Drift @ Ambient Temp.**	NR
8)	**Zero Noise @ Constant Temp.**	NR
9)	**External Power Supply**	NR
10)	**Power Consumption**	NR
11)	**Output Signal**	Millivolt
12)	**Electronic Response Time**	NR
13)	**Operating Temperature Range**	10 to 100°C
14)	**Storage Temperature Range/ Requirements**	NR
15)	**Weight**	NR
16)	**External Dimensions**	NR
17)	**Calibration Requirements/ Maintenance**	Easily replaceable electrode; has gone 6 months without calibration
18)	**Manufacturer**	OMEGA
19)	**Cost**	$105
20)	**Comments**	Self-cleaning, flat surface, combination, ATC element optional, made of CVPC, gel-filled, double-junction, mounting assembly

Global Water pH WQ201 (use with Global Water EZ Sensor Display)

1)	**Analyte**	pH
2)	**Method**	Single-junction combination electrode
3)	**Range/Sensitivity**	0 to 14 pH
4)	**Accuracy**	1% of full scale
5)	**Repeatability/Life Expectancy**	NR
6)	**Zero Drift @ Constant Temp.**	NR
7)	**Zero Drift @ Ambient Temp.**	NR
8)	**Zero Noise @ Constant Temp.**	NR
9)	**External Power Supply**	12 VDC for sensors; 18 VDC for display
10)	**Power Consumption**	NR
11)	**Output Signal**	4 to 20 mA
12)	**Electronic Response Time**	NR
13)	**Operating Temperature Range**	-40 to 55°C
14)	**Storage Temperature Range/ Requirements**	NR
15)	**Weight**	1 lb
16)	**External Dimensions**	0.8" x 8"
17)	**Calibration Requirements/ Maintenance**	6 months to 1 yr
18)	**Manufacturer**	Global Water
19)	**Cost**	$440 for sensor; $285 for LCD display
20)	**Comments**	Rugged; industrial-grade; integral thermistor for temperature compensation; built-in reference electrode; compression fitting available; often put into ground and taken out only every 6 months for cleaning

GLI International Model 6622/2622

1)	**Analyte**	pH
2)	**Method**	Combination electrode
3)	**Range/Sensitivity**	0 to 14 pH
4)	**Accuracy**	Stability—less than 0.05 pH per 24 hr
5)	**Repeatability/Life Expectancy**	NR
6)	**Zero Drift @ Constant Temp.**	NR
7)	**Zero Drift @ Ambient Temp.**	NR
8)	**Zero Noise @ Constant Temp.**	NR
9)	**External Power Supply**	NR
10)	**Power Consumption**	NR
11)	**Output Signal**	NR
12)	**Electronic Response Time**	NR
13)	**Operating Temperature Range**	0 to 80°C (32 to 176°F)
14)	**Storage Temperature Range/ Requirements**	NR
15)	**Weight**	NR
16)	**External Dimensions**	5.9" x 0.875" diameter
17)	**Calibration Requirements/ Maintenance**	NR
18)	**Manufacturer**	GLI International, Inc.
19)	**Cost**	$135
20)	**Comments**	Maximum flowrate of 10 ft (3 m)/sec, available with temperature compensation (300 ohm thermistor or Pt1000 ohm resistance temperature detector [RTD])

Innovative Sensors Model 10

1)	**Analyte**	pH
2)	**Method**	Single-junction
3)	**Range/Sensitivity**	NR
4)	**Accuracy**	NR
5)	**Repeatability/Life Expectancy**	NR
6)	**Zero Drift @ Constant Temp.**	<1 millivolt (0.03 pH)/day
7)	**Zero Drift @ Ambient Temp.**	NR
8)	**Zero Noise @ Constant Temp.**	NR
9)	**External Power Supply**	Preamplifier requires 9-V battery
10)	**Power Consumption**	NR
11)	**Output Signal**	Millivolt (60 mV/pH unit—positive from 0 to 7; negative from 7 to 14)
12)	**Electronic Response Time**	NR
13)	**Operating Temperature Range**	Up to 80°C
14)	**Storage Temperature Range/ Requirements**	NR
15)	**Weight**	NR
16)	**External Dimensions**	5" in length (front-threaded 0.5" from bottom and rear-threaded at the top); 0.5" NPTM
17)	**Calibration Requirements/ Maintenance**	Recommended on a monthly basis
18)	**Manufacturer**	Innovative Sensors
19)	**Cost**	$70 for probe; $75 for preamplifier (one time)
20)	**Comments**	No ATC

Rosemount 399 (based on use with Rosemount 1054B CL Analyzer)

1)	**Analyte**	pH
2)	**Method**	Flat glass, self-cleaning, disposable electrode
3)	**Range/Sensitivity**	0 to 14 pH; resolution of 0.01 pH unit
4)	**Accuracy**	NR
5)	**Repeatability/Life Expectancy**	0.25% of span over 30 days
6)	**Zero Drift @ Constant Temp.**	NR
7)	**Zero Drift @ Ambient Temp.**	NR
8)	**Zero Noise @ Constant Temp.**	NR
9)	**External Power Supply**	115 VAC ±10%, 50/60 Hz ±6%, 4.0 W or 230 VAC ±10%, 50/60 Hz ±6%, 4.0 W for analyzer
10)	**Power Consumption**	NR
11)	**Output Signal**	mV (transmission capability up to 3 mi) from sensor; 0 to 20 mA or 4 to 20 mA into 600 ohms maximum load for analyzer
12)	**Electronic Response Time**	NR
13)	**Operating Temperature Range**	−10 to 65°C for analyzer
14)	**Storage Temperature Range/ Requirements**	NR
15)	**Weight**	0.45 kg (1 lb)
16)	**External Dimensions**	5.67" L x 1" diameter
17)	**Calibration Requirements/ Maintenance**	NR
18)	**Manufacturer**	Rosemount
19)	**Cost**	$291
20)	**Comments**	Combination electrode with integral preamplifier; automatic temperature compensation from 0 to 85°C (32 to 185°F)

Rosemount Model 499A CL Sensor with Model 1054B Analyzer

1) Analyte	Free residual chlorine (should be used with pH sensor — refer to Rosemount 399 pH sensor)
2) Method	Platinum cathode and silver anode; electrochemical reaction results in current flow proportional to amount of chlorine
3) Range/Sensitivity	0 to 20 ppm (mg/L) chlorine, 0 to 14 pH for analyzer; 32 to 122°F (0 to 50°C)
4) Accuracy	Analyzer resolution is 0.001 ppm free residual chlorine (as Cl_2) and 0.01 pH unit
5) Repeatability/Life Expectancy	Analyzer output stability ±0.25% of span over 30 days
6) Zero Drift @ Constant Temp.	NR
7) Zero Drift @ Ambient Temp.	NR
8) Zero Noise @ Constant Temp.	NR
9) External Power Supply	115 VAC, ±10%, 50/60 Hz ±6%, 4.0 W; 230 VAC, ±10%, 50/60 Hz ±6%, 4.0 W
10) Power Consumption	NR
11) Output Signal	Isolated, 0 to 20 mA or 4 to 20 mA into 600 ohms maximum load; direct or reverse acting; dampening: 0 to 255 sec; output 1: total free chlorine (ppm); output 2 (optional): pH
12) Electronic Response Time	NR
13) Operating Temperature Range	-10 to 65°C (14 to 149°F)
14) Storage Temp. Range/ Requirements	NR
15) Weight	Analyzer is 1.1 kg (2.5 lb)
16) External Dimensions	Sensor is 5.63" in length and 1" diameter; analyzer is 144 x 144 x 192 mm (5.7 x 5.7 x 7.6 in)
17) Calibration Requirements/ Maintenance	Two-point buffer pH calibration standard for automatic pH correction
18) Manufacturer	Rosemount
19) Cost	$1,497 for microprocessor analyzer; $663 for chlorine sensor; $291 for disposable pH sensor (see template for Rosemount 399 pH sensor)
20) Comments	Automatic temperature correction factor from 0 to 50°; if the pH is greater than 7 and fluctuates by more than 0.1 unit, continuous measurement of pH and automatic pH correction are necessary; automatic pH correction from 5.0 to 9.5; alarm system with dual setpoint

GLI International Model 95 CL Residual Chlorine Monitoring System

1) Analyte	Free residual chlorine, hypochlorous acid (HOCl), pH, temperature
2) Method	Amperometric, dual electrode
3) Range/Sensitivity	0.00 to 5.00 ppm for free residual chlorine, 0.00 to 5.00 ppm for HOCl, 0.00 to 14.00 for pH, 0.0 to 100.0°C for temperature; sensitivity is 0.01 pH, 0.05 ppm, and 0.1°C
4) Accuracy	0.1 pH, 0.1 ppm, and 0.1°C
5) Repeatability/Life Expectancy	0.01 pH, 0.05 ppm, and 0.05°C
6) Zero Drift @ Constant Temp.	0.008% of span per °C
7) Zero Drift @ Ambient Temp.	NR
8) Zero Noise @ Constant Temp.	NR
9) External Power Supply	98 to 132 VAC, 50/60 Hz (less than 15 VA); optional 205 to 275 VAC, 50/60 Hz
10) Power Consumption	NR
11) Output Signal	Four 0 to 20 mA or 4 to 20 mA outputs can be independently set up (500 ohms maximum load)
12) Electronic Response Time	30 sec typical, 2 min full scale
13) Operating Temperature Range	Thermal temperature equilibrium up to 1 hour (approximately)
14) Storage Temp. Range/ Requirements	−30 to 50°C, 0 to 100% humidity for analyzer; 5 to 50°C for chlorine and pH sensors
15) Weight	Analyzer net weight 12 lb (5.5 kg); sampling assembly net weight 15 lb (6.8 kg)
16) External Dimensions	Analyzer is 12.5 x 9.7 x 6.57"; sampling assembly is 14.5 x 13.75 x 7.63"
17) Calibration Requirements/ Maintenance	Replaceable chlorine sensor membrane cap; buffers and reagents not required
18) Manufacturer	GLI International, Inc.
19) Cost	$1,550 for analyzer; $2,565 for sampling assembly
20) Comments	Temperature and pH compensated; for applications when pH does not exceed 8.5 (buffering system available if pH exceeds 8.5); preplumbed sampling assembly; solid-state relays provide dedicated control and alarm functions for free residual chlorine and pH; automatic temperature compensation from 5 to 50°C

Hanna Instruments PCA 300

1) Analyte	Free chlorine
2) Method	Adaptation of DPD 330.5; chemical reaction causes color with intensity that is proportional to chlorine concentration; color intensity is measured photometrically
3) Range/Sensitivity	0 to 5 mg/L range; 0.01 mg/L resolution; 0.05 mg/L minimum detectable level
4) Accuracy	±5% of reading or ±0.05 mg/L, whichever is greater
5) Repeatability/Life Expectancy	±0.05 mg/L
6) Zero Drift @ Constant Temp.	NR
7) Zero Drift @ Ambient Temp.	NR
8) Zero Noise @ Constant Temp.	NR
9) External Power Supply	20 VA at 115 VAC/230 VAC; 50/60 Hz
10) Power Consumption	NR
11) Output Signal	Selectable: 0 to 10 mV, 0 to 100 mV, 0 to 1 V or 4 to 20 mA. Output span can be set anywhere in the 0 to 5 mg/L range
12) Electronic Response Time	Depends on the sampling time selected. Typical for a full-scale step change and 2.5 min between 2 consecutive samples: 1 sampling cycle for 90% response and 2 sampling cycles for 100% response.
13) Operating Temperature Range	32 to 104°F (0 to 40°C); sample temperature range of 41 to 104°F (5 to 40°C)
14) Storage Temp. Range/ Requirements	NR
15) Weight	11 lb (5 kg) without reagents
16) External Dimensions	12.5 x 10.5 x 6.25" (318 x 267 x 159 mm)
17) Calibration Requirements/ Maintenance	Reagents include HI 70450 indicator, HI 70451 buffer, HI 70452 DPD compound (life is 50 days based on 10-min sampling rate); no calibration is necessary for PCA models, however it can be done with the built-in Calibration Check System if required for regulatory purposes
18) Manufacturer	Hanna Instruments, Inc. (HI)
19) Cost	$2,150
20) Comments	Sampling rate adjustable from 2.5 to 99 min per sample; minimum and maximum flowrates are 100 mL/min and 500 mL/min respectively; hardness must not exceed 1,000 mg/L as $CaCO_3$; two level setpoints for process control

ATI Series A15

1) Analyte	Free chlorine; can be configured to monitor chloramine if a separate sensor is purchased
2) Method	Polarographic membrane sensor; pair of electrodes immersed in a conductive electrolyte and isolated from the sample by a chlorine-permeable membrane; chlorine is reduced to chloride on the surface of the electrode
3) Range/Sensitivity	0 to 2.000 and 0 to 20.00 ppm standard, 0 to 200.0 ppm optional with jumper; sensitivity is 0.001 ppm above 0.020 ppm
4) Accuracy	Linearity is 0.5%
5) Repeatability/Life Expectancy	0.05 ppm
6) Zero Drift @ Constant Temp.	NR
7) Zero Drift @ Ambient Temp.	NR
8) Zero Noise @ Constant Temp.	NR
9) External Power Supply	120/230 VAC, 50/60 Hz, 5 VA maximum
10) Power Consumption	NR
11) Output Signal	Isolated 4 to 20 mA, 600 ohms maximum; programmable for any range from 0 to 1 ppm to 0 to 20 ppm
12) Electronic Response Time	90% in <60 sec
13) Operating Temperature Range	Electronics −20 to +50°C, sensor 0 to 50°C; 0 to 99% noncondensing humidity
14) Storage Temp. Range/ Requirements	NR
15) Weight	<20 lb
16) External Dimensions	Probe is 5.5" long and 2.5" in diameter; monitor is 3.5 x 3.5 x 5.5"; flow cell is 7.5 x 4 x 3"
17) Calibration Requirements/ Maintenance	Does not require chemical addition
18) Manufacturer	ATI
19) Cost	$2,525 for free chlorine; $2,350 for chloramine configuration
20) Comments	Control and alarm relays; automatic temperature compensation from −2 to +52°C; sensor made of Noryl and stainless steel; 25-ft standard interconnect cable; 7 gph minimum flowrate; must have flow cell for free chlorine monitoring

Hach CL17

1) Analyte	Free or total residual chlorine (change from free to total by adjusting reagents)
2) Method	DPD colorimetric method (based on Method 408.E in ASTM *Standard Methods*)
3) Range/Sensitivity	0 to 5 mg/L range (free or total residual chlorine); 0.01 mg/L resolution; minimum detection 0.05 mg/L
4) Accuracy	Better than ±5% of reading or ±0.05 mg/L, whichever is greater
5) Repeatability/Life Expectancy	Within ±0.05 mg/L
6) Zero Drift @ Constant Temp.	NR
7) Zero Drift @ Ambient Temp.	NR
8) Zero Noise @ Constant Temp.	NR
9) External Power Supply	125 VAC, 15 A; or 230 VAC, 10 A
10) Power Consumption	NR
11) Output Signal	0 to 10 mV, 0 to 100 mV, 0 to 1V or 4 to 20 mA; output span programmable over any portion of the 0 to 5 mg/L range
12) Electronic Response Time	NR
13) Operating Temperature Range	NR
14) Storage Temp. Range/ Requirements	NR
15) Weight	31 lb
16) External Dimensions	302 mm W x 371 mm H x 175 mm D
17) Calibration Requirements/ Maintenance	Low maintenance, 473 mL each of indicator and buffer per month; pump module tubing to be replaced at 3- to 6-month intervals
18) Manufacturer	Hach
19) Cost	$2,375
20) Comments	Performs complete analysis every 2.5 min; automatic color/ turbidity compensation; two sample concentration alarms, analyzer system warning and analyzer system shutdown alarms

Orion Residual Chlorine Electrode, Combination

1) Analyte	Total chlorine
2) Method	Based on ASTM iodometric method, adding iodide to sample to form free iodine
3) Range/Sensitivity	0.01 to 20 ppm
4) Accuracy	NR
5) Repeatability/Life Expectancy	±2% of value between 0.05 and 50 ppm
6) Zero Drift @ Constant Temp.	NR
7) Zero Drift @ Ambient Temp.	NR
8) Zero Noise @ Constant Temp.	NR
9) External Power Supply	NR
10) Power Consumption	NR
11) Output Signal	NR
12) Electronic Response Time	95% in 10 sec
13) Operating Temperature Range	0 to 50°C
14) Storage Temp. Range/ Requirements	NR
15) Weight	NR
16) External Dimensions	13.9-cm-long by 12-mm-diameter body (cap diameter 16 mm) and a 75-cm cable
17) Calibration Requirements/ Maintenance	Requires iodide source
18) Manufacturer	Orion
19) Cost	$389
20) Comments	NR

Orion Model 1770

1)	**Analyte**	Total residual chlorine monitor
2)	**Method**	Potentiometric determination of iodine produced by the reaction of chlorine with iodide
3)	**Range/Sensitivity**	0.001 to 10 ppm chlorine (other ranges available upon request); error less than ±10% of reading
4)	**Accuracy**	Detection limit of 0.001 ppm chlorine
5)	**Repeatability/Life Expectancy**	Better than ±5% of reading
6)	**Zero Drift @ Constant Temp.**	NR
7)	**Zero Drift @ Ambient Temp.**	NR
8)	**Zero Noise @ Constant Temp.**	NR
9)	**External Power Supply**	100/115/220 VAC, 50/60 Hz, 230 watts
10)	**Power Consumption**	NR
11)	**Output Signal**	4 to 20 mA (300 ohms maximum load) and 0 to 5 VDC (10 K ohms minimum load)
12)	**Electronic Response Time**	NR
13)	**Operating Temperature Range**	0 to 45°C
14)	**Storage Temp. Range/ Requirements**	0 to 45°C
15)	**Weight**	160 lb
16)	**External Dimensions**	60" H, 25" W, 13.5" D
17)	**Calibration Requirements/ Maintenance**	Two-point, chemical calibration; reagent consumption is 20 L/month (5 gal/month) and is supplied as two-part concentrate
18)	**Manufacturer**	Orion
19)	**Cost**	$9,825
20)	**Comments**	Requires flow of at least 2 L/min (0.5 gpm) and alkalinity less than 500 ppm; wall mounting

Barnant Thermocouple

1)	**Analyte**	Temperature
2)	**Method**	Bimetallic conduit, iron-constantan
3)	**Range/Sensitivity**	-31 to 200°F
4)	**Accuracy**	2% of full scale or 0.33% of reading
5)	**Repeatability/Life Expectancy**	NR
6)	**Zero Drift @ Constant Temp.**	NR
7)	**Zero Drift @ Ambient Temp.**	NR
8)	**Zero Noise @ Constant Temp.**	NR
9)	**External Power Supply**	Voltage, 12 VDC loop
10)	**Power Consumption**	NR
11)	**Output Signal**	Millivolt proportional to temperature
12)	**Electronic Response Time**	Continuous sensing after 9-sec initial response time
13)	**Operating Temperature Range**	NR
14)	**Storage Temperature Range/ Requirements**	NR
15)	**Weight**	<0.5 lb
16)	**External Dimensions**	0.118" D x 4.0" immersion depth
17)	**Calibration Requirements/ Maintenance**	Factory calibration, field confirmation
18)	**Manufacturer**	Barnant
19)	**Cost**	Catalog #685-1771, $30
20)	**Comments**	Industrial-grade sensor, direct insertion or immersion into water, comes with 6' cable

Davis Instruments

1) **Analyte**	Temperature
2) **Method**	Thermocouple T-type, copper-constantan, self-adhesive surface patch
3) **Range/Sensitivity**	−75 to 200°F
4) **Accuracy**	±1.5°F
5) **Repeatability/Life Expectancy**	NR
6) **Zero Drift @ Constant Temp.**	Better than 2 μV/°C referred to input
7) **Zero Drift @ Ambient Temp.**	NR
8) **Zero Noise @ Constant Temp.**	NR
9) **External Power Supply**	8 to 40 VDC for the transmitter
10) **Power Consumption**	NR
11) **Output Signal**	4 to 20 mA
12) **Electronic Response Time**	Continuous sensing
13) **Operating Temperature Range**	0 to 60°C for the transmitter
14) **Storage Temperature Range/ Requirements**	NR
15) **Weight**	<0.25 lb
16) **External Dimensions**	NR
17) **Calibration Requirements/ Maintenance**	Field calibration to determine current linearity of response to output of thermocouple
18) **Manufacturer**	Davis Instruments
19) **Cost**	Catalog #ST21M, $75; two-wire thermocouple transmitter, $150
20) **Comments**	Nonintrusive surface patch thermocouple

Global Water

1)	**Analyte**	Temperature Model WQ101
2)	**Method**	Bimetallic conduit
3)	**Range/Sensitivity**	−50 to 50°C
4)	**Accuracy**	±0.1°C
5)	**Repeatability/Life Expectancy**	NR
6)	**Zero Drift @ Constant Temp.**	NR
7)	**Zero Drift @ Ambient Temp.**	NR
8)	**Zero Noise @ Constant Temp.**	NR
9)	**External Power Supply**	12 VDC
10)	**Power Consumption**	NR
11)	**Output Signal**	4 to 20 mA linear
12)	**Electronic Response Time**	Continuous sensing
13)	**Operating Temperature Range**	−40 to 55°C
14)	**Storage Temperature Range/ Requirements**	NR
15)	**Weight**	0.5 lb
16)	**External Dimensions**	0.625" D x 8" L
17)	**Calibration Requirements/ Maintenance**	Factory calibration, field confirmation
18)	**Manufacturer**	Global Water
19)	**Cost**	$200
20)	**Comments**	Industrial-grade sensor, direct insertion or immersion into water, comes with 25' cable

OMEGA GJQIN-18-12

1)	**Analyte**	Temperature
2)	**Method**	Thermocouple J-type
3)	**Range/Sensitivity**	−250 to 750°C
4)	**Accuracy**	±0.4% full scale; resolution of 1°F or 1°C
5)	**Repeatability/Life Expectancy**	NR
6)	**Zero Drift @ Constant Temp.**	NR
7)	**Zero Drift @ Ambient Temp.**	NR
8)	**Zero Noise @ Constant Temp.**	NR
9)	**External Power Supply**	Signal conditioning module, 4.5 VDC
10)	**Power Consumption**	NR
11)	**Output Signal**	Millivolt signal proportional to temperature
12)	**Electronic Response Time**	Continuous sensing
13)	**Operating Temperature Range**	−10 to 60°C
14)	**Storage Temperature Range/ Requirements**	NR
15)	**Weight**	Conditioning module, 10 oz
16)	**External Dimensions**	5.71" H x 3.15" W x 1.44" D
17)	**Calibration Requirements/ Maintenance**	Factory calibration, field confirmation
18)	**Manufacturer**	OMEGA
19)	**Cost**	Probe GJQIN-18-12, $30; MOD-05/16, $142
20)	**Comments**	Industrial-grade sensor, direct insertion or immersion into water, or placed into thermowell

OMEGA ON-970-44004

1)	**Analyte**	Temperature
2)	**Method**	Thermistor
3)	**Range/Sensitivity**	−40 to 100°C
4)	**Accuracy**	±0.3% FS; resolution of 0.1°C
5)	**Repeatability/Life Expectancy**	NR
6)	**Zero Drift @ Constant Temp.**	NR
7)	**Zero Drift @ Ambient Temp.**	NR
8)	**Zero Noise @ Constant Temp.**	NR
9)	**External Power Supply**	Signal conditioning module, 4.5 VDC
10)	**Power Consumption**	NR
11)	**Output Signal**	Millivolt signal proportional to temperature
12)	**Electronic Response Time**	Continuous sensing
13)	**Operating Temperature Range**	−10 to 60°C
14)	**Storage Temperature Range/ Requirements**	NR
15)	**Weight**	Conditioning module, 10 oz
16)	**External Dimensions**	5.71" H x 3.15" W x 1.44" D
17)	**Calibration Requirements/ Maintenance**	Factory calibration, field confirmation
18)	**Manufacturer**	OMEGA
19)	**Cost**	Probe ON-970-44004, $55; MOD-03, $101
20)	**Comments**	Industrial-grade sensor, direct insertion into water through the pipe via ⅛" NPT

OMEGA RTD-810

1) Analyte	Temperature
2) Method	Resistance temperature device
3) Range/Sensitivity	−200 to 750°C
4) Accuracy	±0.3% FS; resolution of 0.1°F or 0.1°C
5) Repeatability/Life Expectancy	NR
6) Zero Drift @ Constant Temp.	NR
7) Zero Drift @ Ambient Temp.	NR
8) Zero Noise @ Constant Temp.	NR
9) External Power Supply	Signal conditioning module, 4.5 VDC
10) Power Consumption	NR
11) Output Signal	Millivolt signal proportional to temperature
12) Electronic Response Time	Continuous sensing
13) Operating Temperature Range	−10 to 60°C
14) Storage Temperature Range/ Requirements	NR
15) Weight	Conditioning module, 10 oz
16) External Dimensions	5.71" H x 3.15" W x 1.44" D
17) Calibration Requirements/ Maintenance	Factory calibration, field confirmation
18) Manufacturer	OMEGA
19) Cost	Probe RTD-810, $79; MOD-08/09, $142
20) Comments	Industrial-grade sensor, direct insertion into water through the pipe via $^{1}/_{8}$" NPT

OMEGA OS-614

1) Analyte	Temperature
2) Method	Infrared pyrometer
3) Range/Sensitivity	0 to 250°F
4) Accuracy	±1°F
5) Repeatability/Life Expectancy	Repeatability of ±0.5% of reading ±1 digit
6) Zero Drift @ Constant Temp.	NR
7) Zero Drift @ Ambient Temp.	NR
8) Zero Noise @ Constant Temp.	NR
9) External Power Supply	9 V
10) Power Consumption	NR
11) Output Signal	Analog output of 1 mV per degree
12) Electronic Response Time	Continuous sensing
13) Operating Temperature Range	32 to 120°F
14) Storage Temperature Range/ Requirements	NR
15) Weight	Probe, 5 oz; display unit, 9.6 oz
16) External Dimensions	Probe, 1.3" W x 1.3" H x 6.2" D; Display, 3.2" W x 1.3" H x 6.2" D
17) Calibration Requirements/ Maintenance	Factory calibration, field confirmation
18) Manufacturer	OMEGA
19) Cost	Model OS-614, $650
20) Comments	Noncontact measurements, minimum target size of 0.25" D at 0.75" distance; distance to target ratio of 3:1

Sievers 800 TOC Analyzer

1)	**Analyte**	TOC
2)	**Method**	Selective membrane/conductivity detection cell and strong UV/persulfate oxidation
3)	**Range/Sensitivity**	0.05 ppb to 50,000 ppb in hot or cold water
4)	**Accuracy**	NR
5)	**Repeatability/Life Expectancy**	±3% of total carbon value
6)	**Zero Drift @ Constant Temp.**	NR
7)	**Zero Drift @ Ambient Temp.**	NR
8)	**Zero Noise @ Constant Temp.**	NR
9)	**External Power Supply**	Model 800—110 V, 60 Hz, 50 W; Model 810—100V, 50/60 Hz, 50 W; Model 820—220 V, 50 Hz, 50 W
10)	**Power Consumption**	NR
11)	**Output Signal**	0 to 10 V, 2 to 20 mA, RS-232, parallel printer port
12)	**Electronic Response Time**	6-min analysis time
13)	**Operating Temperature Range**	0 to 100°C water
14)	**Storage Temp. Range/ Requirements**	10 to 40°C (50 to 104°F)
15)	**Weight**	30 lb (13.5 kg)
16)	**External Dimensions**	6" H x 19.5" W x 13.5" D (16.5 x 50 x 34 cm)
17)	**Calibration Requirements/ Maintenance**	Low maintenance costs; quarterly replacement of prepackaged, disposable reagent modules typically takes 5 min; semiannual UV lamp and annual pump tubing replacement takes less than 10 min; excellent stability for up to 1 yr; easy-to-verify calibration using standard solutions; prepackaged 3-month supply of reagents (persulfate and phosphoric acid)
18)	**Manufacturer**	Sievers
19)	**Cost**	$19,800 for analyzer; $150/box for acid and oxidizer (~3-month supply but depends on application)
20)	**Comments**	User-friendly; easy data transfer to PC, self-diagnostics, remote alarms and controls

Maihak TOCOR 20

1) Analyte	TOC
2) Method	Thermal oxidation of organic carbon and subsequent detection of CO_2 in treated gas stream
3) Range/Sensitivity	0 to 50 mg/L; 0 to 100 mg/L; 0 to 200 mg/L; 0 to 500 mg/L; and 0 to 1,000 mg/L; detection limit <1% of measuring range
4) Accuracy	NR
5) Repeatability/Life Expectancy	NR
6) Zero Drift @ Constant Temp.	Zero drift and span drift are <3%/wk
7) Zero Drift @ Ambient Temp.	NR
8) Zero Noise @ Constant Temp.	NR
9) External Power Supply	230 VAC, 50/60 Hz (+10/−1500 VA)
10) Power Consumption	NR
11) Output Signal	4 to 20 mA
12) Electronic Response Time	NR
13) Operating Temperature Range	NR
14) Storage Temp. Range/ Requirements	5 to 30°C ambient temperature
15) Weight	Approximately 70 kg
16) External Dimensions	1200 x 600 x 500 mm
17) Calibration Requirements/ Maintenance	NR
18) Manufacturer	Maihak
19) Cost	Analog unit, $19,000
20) Comments	Optional dilution system and backflush filter

Tytronics® Sentinel Manganese Analyzer

1) Analyte	Manganese
2) Method	Online colorimetric method
3) Range/Sensitivity	0 to 200 ppb
4) Accuracy	NR
5) Repeatability/Life Expectancy	NR
6) Zero Drift @ Constant Temp.	NR
7) Zero Drift @ Ambient Temp.	NR
8) Zero Noise @ Constant Temp.	NR
9) External Power Supply	NR
10) Power Consumption	NR
11) Output Signal	4 to 20 mA
12) Electronic Response Time	Analysis time is approximately 10 min
13) Operating Temperature Range	NR
14) Storage Temperature Range/ Requirements	NR
15) Weight	NR
16) External Dimensions	NR
17) Calibration Requirements/ Maintenance	Reagents include pH buffer and formaldoxime; three calibration methods
18) Manufacturer	Tytronics®
19) Cost	$8,950 for single stream
20) Comments	Batch sampling minimizes reagent consumption; alarm function

Tytronics® Sentinel Iron Analyzer (FPA 806)

1) Analyte	Iron
2) Method	Online colorimetric method; immersion fiber optics reflectance probe
3) Range/Sensitivity	0 to 500 ppb; 0 to 1 ppm; 0 to 10 ppm; 0 to 100 ppm
4) Accuracy	NR
5) Repeatability/Life Expectancy	NR
6) Zero Drift @ Constant Temp.	NR
7) Zero Drift @ Ambient Temp.	NR
8) Zero Noise @ Constant Temp.	NR
9) External Power Supply	NR
10) Power Consumption	NR
11) Output Signal	4 to 20 mA
12) Electronic Response Time	Analysis time is approximately 10 min
13) Operating Temperature Range	NR
14) Storage Temperature Range/ Requirements	NR
15) Weight	NR
16) External Dimensions	NR
17) Calibration Requirements/ Maintenance	Reagents include ferrozine/hydroxylamine and acetate buffer
18) Manufacturer	Tytronics®
19) Cost	$8,950 for single stream
20) Comments	Self-diagnostics

Rosemount Model Chempure™

1)	**Analyte**	Iron
2)	**Method**	Online colorimetric method
3)	**Range/Sensitivity**	0 to 300 ppb, 0 to 1 ppm, or 0 to 10 ppm; detection limit is ±1% of full scale
4)	**Accuracy**	NR
5)	**Repeatability/Life Expectancy**	NR
6)	**Zero Drift @ Constant Temp.**	NR
7)	**Zero Drift @ Ambient Temp.**	NR
8)	**Zero Noise @ Constant Temp.**	NR
9)	**External Power Supply**	115 VAC, 60 Hz, 3 amp
10)	**Power Consumption**	NR
11)	**Output Signal**	0 to 5 VDC; 4 to 20 mA
12)	**Electronic Response Time**	NR
13)	**Operating Temperature Range**	0 to 50°C (32 to 122°F) sample temperature
14)	**Storage Temperature Range/ Requirements**	10 to 40°C (50 to 104°F) ambient temperature; temperature swings greater than ±10°F not recommended
15)	**Weight**	Approximately 126 lb
16)	**External Dimensions**	44 x 16 x 18"
17)	**Calibration Requirements/ Maintenance**	15 min monthly for reagent and tubing replacement; automatic calibration
18)	**Manufacturer**	Rosemount Analytical
19)	**Cost**	$13,778 for analyzer; $2,255 for 1-yr reagent supply
20)	**Comments**	Self-diagnostics

Capital Controls Aztec® M100

1)	**Analyte**	Manganese
2)	**Method**	Formaldoxime colorimetric method; based on reaction of formaldoxime in an alkaline sample; light passes through a 34-mm sample cell at a wavelength of 450 nm
3)	**Range/Sensitivity**	20 to 500 ppb Mn^{2+}; resolution is 0.01 mg/L
4)	**Accuracy**	5% at full range
5)	**Repeatability/Life Expectancy**	<2% of reading
6)	**Zero Drift @ Constant Temp.**	NR
7)	**Zero Drift @ Ambient Temp.**	NR
8)	**Zero Noise @ Constant Temp.**	NR
9)	**External Power Supply**	110 VAC, ±6%, 60 Hz, 120/240 VAC, ±6%, 50/60 Hz
10)	**Power Consumption**	240 W
11)	**Output Signal**	Isolated 4 to 20 mADC into 1000 ohms maximum; RS232
12)	**Electronic Response Time**	NR
13)	**Operating Temperature Range**	0 to 35°C (32 to 96°F) operating range; 1 to 35°C (33 to 96°F) sample range
14)	**Storage Temperature Range/ Requirements**	NR
15)	**Weight**	125 lb (60 kg) shipping weight
16)	**External Dimensions**	NR
17)	**Calibration Requirements/ Maintenance**	Two-point autocalibration based on deionized water and a manganese standard; operates a minimum of 20 days without reagent replacement; chemical reagents include alkali reagent, complexing reagent, color reagent (formaldoxime), deionized water, and standard solution
18)	**Manufacturer**	Capital Controls Aztec
19)	**Cost**	$18,000 for monitor and mounting; $380/month for reagents
20)	**Comments**	Self-cleaning optics; rugged industrial design; programmable sample frequency; self-diagnostics; microprocessor controlled; 200 to 500 mL/min sample flow; alarm system

Capital Controls Aztec FE100

1) Analyte	Iron
2) Method	TPTZ colorimetric method; based on the reaction of 2,4,6-tripyridyl-*s*-triazine in an acidic sample; light passes through a 20-mm sample cell at a wavelength of 590 nm
3) Range/Sensitivity	0.020 to 2.00 mg/L
4) Accuracy	±5%
5) Repeatability/Life Expectancy	<2% of reading
6) Zero Drift @ Constant Temp.	NR
7) Zero Drift @ Ambient Temp.	NR
8) Zero Noise @ Constant Temp.	NR
9) External Power Supply	110 VAC, ±6%, 60 Hz, 120/240 VAC, ±6%, 50/60 Hz
10) Power Consumption	240 W
11) Output Signal	Isolated 4 to 20 mADC; serial (RS232) and parallel (centronics)
12) Electronic Response Time	NR
13) Operating Temperature Range	0 to 35°C (32 to 96°F) operating range; 1 to 35°C (33 to 96°F) sample range
14) Storage Temperature Range/ Requirements	NR
15) Weight	125 lb (60 kg) shipping weight
16) External Dimensions	NR
17) Calibration Requirements/ Maintenance	Two-point autocalibration based on deionized water and an iron standard; self-cleaning optics; operates a minimum of 20 days without reagent replacement; chemical reagents include acid reagent, buffer reagent, color reagent (TPTZ), deionized water, and standard solution
18) Manufacturer	Capital Controls Aztec
19) Cost	$18,000 for monitor and mounting; $186/month for reagents
20) Comments	Self-diagnostics; microprocessor controlled; programmable sample frequency; sample flow of 200 to 500 mL/min; alarm system

Hach Model 31800

1) Analyte	Manganese
2) Method	DPD pump-colorimeter
3) Range/Sensitivity	0 to 0.2 mg/L
4) Accuracy	±5% of range at constant temperature
5) Repeatability/Life Expectancy	±1% of range at constant temperature
6) Zero Drift @ Constant Temp.	Less than 1% of range
7) Zero Drift @ Ambient Temp.	Less than 1% of range
8) Zero Noise @ Constant Temp.	NR
9) External Power Supply	98 to 130 VAC, 60 or 50 Hz, 40 W maximum
10) Power Consumption	NR
11) Output Signal	Voltage outputs of 0 to 10 mV, 0 to 100 mV, 0 to 1 V, 0 to 10 V and current outputs of 0 to 20 mA, 10 to 50 mA, 10 to 50 mA
12) Electronic Response Time	For a full-scale step change in concentration, the analyzer detects a change in 1 min; reaches 96% full scale in 2.5 min
13) Operating Temperature Range	10 to 40°C (50 to 104°F) sample temperature
14) Storage Temperature Range/ Requirements	0 to 50°C (32 to 122°F)
15) Weight	29.5 kg (65 lb) shipping weight
16) External Dimensions	41.3 cm W x 51.4 cm H x 30 cm D (16.25" x 20.25" x 11.7")
17) Calibration Requirements/ Maintenance	Uses DPD indicator and buffer solution and calibration standards; weekly standardization and buffer replenishment; reagent consumption is 946 mL (1 qt); replace tubing every 3 to 6 months
18) Manufacturer	Hach
19) Cost	$4,595
20) Comments	Alarm system; oxidizing agents (i.e., iodine, bromine, chlorine, ozone) can be an interference; will be discontinued soon — has not provided good, consistent performance

Detora On-Line Metals Monitoring System

1) Analyte	Metals; sensitive to elements from Al to U (atomic numbers 13-92)
2) Method	Metal particulates accumulate on filters and x-ray fluorescence is used to detect the mass of accumulated metals
3) Range/Sensitivity	Sub-ppb to percent; detection limit of 0.007 to 0.01 ppb (depending on sampling frequency)
4) Accuracy	NR
5) Repeatability/Life Expectancy	Precision is 1.5% of average over calibration range
6) Zero Drift @ Constant Temp.	NR
7) Zero Drift @ Ambient Temp.	NR
8) Zero Noise @ Constant Temp.	NR
9) External Power Supply	110 V, 11 A, 60 Hz
10) Power Consumption	NR
11) Output Signal	NR
12) Electronic Response Time	NR
13) Operating Temperature Range	Sample temperature of 32 to 140°F (0 to 50°C)
14) Storage Temperature Range/ Requirements	NR
15) Weight	NR
16) External Dimensions	25" x 26" footprint
17) Calibration Requirements/ Maintenance	No reagents; no waste; low maintenance; calibration using prepared standards in measurement range; accepts standard 0.45 micron polyvinylidene difluoride (PVDF) filters; filter replacement frequency depends on concentration in stream
18) Manufacturer	Detora
19) Cost	$125,000 for single channel for particulate iron and manganese; $187,000 for dissolved
20) Comments	100 mCi Cm-244 x-ray source; 0.2 to 0.5 L/min user-controllable flowrate

Solomat 603IS Nitrate Electrode with 520C Display

1) Analyte	Nitrate
2) Method	Ion-selective electrode
3) Range/Sensitivity	0.18 to 230 ppm
4) Accuracy	NR
5) Repeatability/Life Expectancy	NR
6) Zero Drift @ Constant Temp.	NR
7) Zero Drift @ Ambient Temp.	NR
8) Zero Noise @ Constant Temp.	NR
9) External Power Supply	110 VAC for display meter
10) Power Consumption	NR
11) Output Signal	mV
12) Electronic Response Time	NR
13) Operating Temperature Range	14 to 122°F (−10 to 50°C)
14) Storage Temperature Range/ Requirements	−5 to 140°F (−20 to 60°C)
15) Weight	1.1 lb for analyzer
16) External Dimensions	Analyzer is 2.9" W x 1.6" deep increasing to 4.4" W x 2.6" deep and is 8.3" long
17) Calibration Requirements/ Maintenance	NR
18) Manufacturer	Solomat
19) Cost	\$749 for electrode; \$239 for reference electrode; \$1,395 for display meter
20) Comments	Bayonet N connector (BNC) plug

OMEGA

1)	**Analyte**	Nitrate
2)	**Method**	Ion-selective combination electrode; double-junction
3)	**Range/Sensitivity**	62,000 to 0.5 ppm for sensor; 0.001 to 1990 for meter
4)	**Accuracy**	±0.5% for meter
5)	**Repeatability/Life Expectancy**	NR
6)	**Zero Drift @ Constant Temp.**	NR
7)	**Zero Drift @ Ambient Temp.**	NR
8)	**Zero Noise @ Constant Temp.**	NR
9)	**External Power Supply**	115 VAC adapter for meter
10)	**Power Consumption**	NR
11)	**Output Signal**	mV
12)	**Electronic Response Time**	30 sec
13)	**Operating Temperature Range**	0 to 50°C
14)	**Storage Temperature Range/ Requirements**	NR
15)	**Weight**	6.4 lb for meter
16)	**External Dimensions**	8.3" x 12" x 4.1" for meter
17)	**Calibration Requirements/ Maintenance**	Two-point automatic calibration
18)	**Manufacturer**	OMEGA
19)	**Cost**	$206 for PVC; $275 for glass
20)	**Comments**	pH range from 2.5 to 11; automatic or manual temperature compensation from 0 to 99.9°C

Cole-Parmer Nitrate

1)	**Analyte**	Nitrate
2)	**Method**	PVC membrane electrode
3)	**Range/Sensitivity**	0.5 to 62,000 ppm for electrode; meter is 0 to 9999 for concentration and ±1999 for mV
4)	**Accuracy**	±0.5% of reading for concentration; ±0.2 for mV
5)	**Repeatability/Life Expectancy**	NR
6)	**Zero Drift @ Constant Temp.**	NR
7)	**Zero Drift @ Ambient Temp.**	NR
8)	**Zero Noise @ Constant Temp.**	NR
9)	**External Power Supply**	110 VAC, 60 Hz; or 220 VAC, 50 Hz
10)	**Power Consumption**	NR
11)	**Output Signal**	±1999 mV, 1 mV/mV unit for meter
12)	**Electronic Response Time**	NR
13)	**Operating Temperature Range**	0 to 50°C
14)	**Storage Temperature Range/ Requirements**	NR
15)	**Weight**	3.0 lb (1.4 kg) shipping weight for meter
16)	**External Dimensions**	9" L x 7" W x 2$^3/_8$" H for meter
17)	**Calibration Requirements/ Maintenance**	NR
18)	**Manufacturer**	Cole-Parmer
19)	**Cost**	$340 for electrode; $795 for meter
20)	**Comments**	Single- or double-junction; pH range is 2.5 to 11; manual or automatic temperature compensation from −5 to 105°C

Orion 930700 (based on use with Orion 290A meter)

1)	**Analyte**	Nitrate
2)	**Method**	Plastic membrane half-cell electrode; direct method
3)	**Range/Sensitivity**	1 to 7 x 10^{-6} M; 14,000 to 0.1 ppm as N; resolution is ±1 least significant digit
4)	**Accuracy**	±0.5% of reading
5)	**Repeatability/Life Expectancy**	NR
6)	**Zero Drift @ Constant Temp.**	NR
7)	**Zero Drift @ Ambient Temp.**	NR
8)	**Zero Noise @ Constant Temp.**	NR
9)	**External Power Supply**	NR
10)	**Power Consumption**	NR
11)	**Output Signal**	NR
12)	**Electronic Response Time**	NR
13)	**Operating Temperature Range**	0 to 40°C for sensor; 5 to 45°C and 5 to 85% relative humidity for meter
14)	**Storage Temperature Range/ Requirements**	Store in 10^{-2} M nitrate standard between samples for short periods; for long-term storage rinse electrode with distilled water, disassemble, and store in vial
15)	**Weight**	NR
16)	**External Dimensions**	5.375" L (13.5 cm)
17)	**Calibration Requirements/ Maintenance**	NR
18)	**Manufacturer**	Orion
19)	**Cost**	$555 for electrode; $690 for meter
20)	**Comments**	PVC gelled membrane; PVC body; available with standard connector or BNC

Orion Ionplus Electrode (based on use with Orion 290A meter)

1)	**Analyte**	Nitrate
2)	**Method**	Plastic membrane combination; direct method
3)	**Range/Sensitivity**	1 to 7 x 10^{-6} M; 14,000 to 0.1 ppm as N; resolution is ±1 least significant digit
4)	**Accuracy**	±0.5% of reading
5)	**Repeatability/Life Expectancy**	NR
6)	**Zero Drift @ Constant Temp.**	NR
7)	**Zero Drift @ Ambient Temp.**	NR
8)	**Zero Noise @ Constant Temp.**	NR
9)	**External Power Supply**	Line adapter to 110 V, 220 V, or 240 V
10)	**Power Consumption**	NR
11)	**Output Signal**	NR
12)	**Electronic Response Time**	NR
13)	**Operating Temperature Range**	0 to 40°C for sensor; 5 to 45°C and 5 to 85% relative humidity for meter
14)	**Storage Temperature Range/ Requirements**	Store in 10^{-2} M nitrate standard between samples for short periods; for long-term storage rinse electrode with distilled water, disassemble, and store in vial
15)	**Weight**	NR
16)	**External Dimensions**	NR
17)	**Calibration Requirements/ Maintenance**	NR
18)	**Manufacturer**	Orion
19)	**Cost**	$625 for electrode; $690 for meter
20)	**Comments**	Membrane replacement recommended every 3 months

Hach One Combination Nitrate Electrode

1) Analyte	Nitrate
2) Method	Combination electrode
3) Range/Sensitivity	0.10 to 14,000 mg/L
4) Accuracy	NR
5) Repeatability/Life Expectancy	NR
6) Zero Drift @ Constant Temp.	NR
7) Zero Drift @ Ambient Temp.	NR
8) Zero Noise @ Constant Temp.	NR
9) External Power Supply	NR
10) Power Consumption	NR
11) Output Signal	NR
12) Electronic Response Time	NR
13) Operating Temperature Range	0 to 40°C
14) Storage Temperature Range/ Requirements	NR
15) Weight	NR
16) External Dimensions	9.4" L x 0.41" diameter
17) Calibration Requirements/ Maintenance	NR
18) Manufacturer	Hach
19) Cost	$315 for electrode
20) Comments	Double-junction; pH range 3 to 11; refillable

Hach APA 6000

1)	**Analyte**	Nitrate
2)	**Method**	Process analyzer with ion-selective electrode
3)	**Range/Sensitivity**	0.1 to 100 mg/L
4)	**Accuracy**	NR
5)	**Repeatability/Life Expectancy**	NR
6)	**Zero Drift @ Constant Temp.**	NR
7)	**Zero Drift @ Ambient Temp.**	NR
8)	**Zero Noise @ Constant Temp.**	NR
9)	**External Power Supply**	NR
10)	**Power Consumption**	NR
11)	**Output Signal**	mA
12)	**Electronic Response Time**	Cycle time less than 5 min
13)	**Operating Temperature Range**	5 to 50°C sample temperature range
14)	**Storage Temperature Range/ Requirements**	NR
15)	**Weight**	NR
16)	**External Dimensions**	NR
17)	**Calibration Requirements/ Maintenance**	Can set up for automatic daily calibration
18)	**Manufacturer**	Hach
19)	**Cost**	$6,200 with AquaTrend interface
20)	**Comments**	Four control loops

Corning Combination (based on use with Corning 355 meter)

1) Analyte	Nitrate
2) Method	Direct; PVC/liquid ion-exchange membrane
3) Range/Sensitivity	1 to 5 x 10^{-6} M; 62,000 to 0.8 ppm
4) Accuracy	±0.5% of ion-selective electrode (ISE) reading
5) Repeatability/Life Expectancy	±1 least significant digit; stability (drift/day) is 0.2% of ISE reading
6) Zero Drift @ Constant Temp.	NR
7) Zero Drift @ Ambient Temp.	NR
8) Zero Noise @ Constant Temp.	NR
9) External Power Supply	NR
10) Power Consumption	NR
11) Output Signal	NR
12) Electronic Response Time	NR
13) Operating Temperature Range	0 to 50°C; pH range of 2.5 to 11
14) Storage Temperature Range/ Requirements	NR
15) Weight	NR
16) External Dimensions	5.5" L (14 cm)
17) Calibration Requirements/ Maintenance	NR
18) Manufacturer	Corning
19) Cost	$450 for electrode; $1,995 for meter
20) Comments	Double-junction reference probe; glass body; meter has automatic temperature compensation from −30 to 130°C

Corning Half-Cell (based on use with Corning 355 meter)

1) Analyte	Nitrate
2) Method	Direct
3) Range/Sensitivity	1 to 5 x 10^{-6}M; 62,000 to 0.8 ppm
4) Accuracy	±0.5% of ISE reading
5) Repeatability/Life Expectancy	±1 least significant digit; stability (drift/day) is 0.2% of ISE reading
6) Zero Drift @ Constant Temp.	NR
7) Zero Drift @ Ambient Temp.	NR
8) Zero Noise @ Constant Temp.	NR
9) External Power Supply	NR
10) Power Consumption	NR
11) Output Signal	NR
12) Electronic Response Time	NR
13) Operating Temperature Range	0 to 50°C; pH range of 2.5 to 11
14) Storage Temperature Range/ Requirements	NR
15) Weight	NR
16) External Dimensions	5.5" L (14 cm)
17) Calibration Requirements/ Maintenance	NR
18) Manufacturer	Corning
19) Cost	$435 for electrode; $1,995 for meter
20) Comments	Meter has automatic temperature compensation from −30 to 130°C

Capital Controls Aztec® N1000 Nitrate Monitor

1) Analyte	Nitrate
2) Method	Ion-selective, nitrate-sensing electrode; sample conditioning with orthophosphate and then heating
3) Range/Sensitivity	Automatic ranging from 0 to 500 mg/L NO_3; resolution up to 0.001 mg/L
4) Accuracy	3% of reading
5) Repeatability/Life Expectancy	NR
6) Zero Drift @ Constant Temp.	NR
7) Zero Drift @ Ambient Temp.	NR
8) Zero Noise @ Constant Temp.	NR
9) External Power Supply	Automatic power recognition from 100 to 250 VAC, 47 to 63 Hz, 1 phase
10) Power Consumption	480 W
11) Output Signal	Single 4 to 20 mADC, 0 to 20 mADC, or 0 to 10 mADC isolated into 1000 ohms maximum
12) Electronic Response Time	90% step change within 5 min
13) Operating Temperature Range	Sample temperature of 32 to 120°F (0 to 50°C); temperature of conditioned sample is 35°C (96°F)
14) Storage Temperature Range/ Requirements	32 to 140°F (0 to 60°C) ambient temperature
15) Weight	35 lb (15 kg) shipping weight
16) External Dimensions	21.5" x 26.5" x 8" (548 mm x 664 mm x 203 mm)
17) Calibration Requirements/ Maintenance	Automatic high and low calibration with nitrate standard solutions; chemical reagents include orthophosphate, low and high nitrate standard solutions, and nitrate probe filling solution
18) Manufacturer	Capital Controls Aztec®
19) Cost	$18,000
20) Comments	Sample stream and electrode are temperature-controlled; sample flow from 200 to 500 mL/min; samples containing particles greater than 100 microns in diameter may require prefiltration

Orion Ionplus Fluoride Electrode

1) **Analyte**	Fluoride
2) **Method**	Solid-state combination electrode; lanthanum fluoride crystal; potential develops across crystal and produces signal based on fluoride concentration
3) **Range/Sensitivity**	Saturated to 0.02 ppm for sensor; 0.000 to 19,900 for meter
4) **Accuracy**	2% error for every degree for sensor; ±0.5% of reading for meter
5) **Repeatability/Life Expectancy**	NR
6) **Zero Drift @ Constant Temp.**	NR
7) **Zero Drift @ Ambient Temp.**	NR
8) **Zero Noise @ Constant Temp.**	NR
9) **External Power Supply**	NR
10) **Power Consumption**	110 V, 220 V, or 240 V for meter
11) **Output Signal**	mV
12) **Electronic Response Time**	NR
13) **Operating Temperature Range**	0 to 80°C
14) **Storage Temperature Range/ Requirements**	5 to 45°C for meter
15) **Weight**	NR
16) **External Dimensions**	140 mm L x 12 mm diameter
17) **Calibration Requirements/ Maintenance**	Recommended daily
18) **Manufacturer**	Orion
19) **Cost**	$555 for electrode; $690 for meter
20) **Comments**	Five-point autocalibration; should mix sample with 50% total ionic strength adjusting buffer (TISAB) to maintain pH of 5

Orion Half-Cell

1) Analyte	Fluoride
2) Method	Half-cell fluoride and reference electrode; lanthanum fluoride crystal; potential develops across crystal and produces signal based on fluoride concentration
3) Range/Sensitivity	Saturated to 0.02 ppm for sensor; 0.000 to 19,900 for meter
4) Accuracy	2% error for every degree for sensor; ±0.5% of reading for meter
5) Repeatability/Life Expectancy	NR
6) Zero Drift @ Constant Temp.	NR
7) Zero Drift @ Ambient Temp.	NR
8) Zero Noise @ Constant Temp.	NR
9) External Power Supply	NR
10) Power Consumption	110 V, 220 V, or 240 V for meter
11) Output Signal	mV
12) Electronic Response Time	NR
13) Operating Temperature Range	0 to 80°C
14) Storage Temperature Range/ Requirements	5 to 45°C for meter
15) Weight	NR
16) External Dimensions	NR
17) Calibration Requirements/ Maintenance	Recommended daily
18) Manufacturer	Orion
19) Cost	$450 for half-cell fluoride electrode; $140 for reference electrode; $690 for meter
20) Comments	Five-point autocalibration; should mix sample with 50% TISAB to maintain pH of 5

Corning Combination (based on use with Corning 355 meter)

1) Analyte	Fluoride
2) Method	Combination electrode; ion-sensitive membrane of crystalline material
3) Range/Sensitivity	1 to 5 x 10^{-6} M; 19,000 to 0.019 ppm
4) Accuracy	±0.5% of ISE reading
5) Repeatability/Life Expectancy	±1 least significant digit; stability (drift/day) is 0.2% of ISE reading
6) Zero Drift @ Constant Temp.	NR
7) Zero Drift @ Ambient Temp.	NR
8) Zero Noise @ Constant Temp.	NR
9) External Power Supply	NR
10) Power Consumption	NR
11) Output Signal	NR
12) Electronic Response Time	NR
13) Operating Temperature Range	0 to 80°; pH range of 5 to 7
14) Storage Temperature Range/ Requirements	NR
15) Weight	NR
16) External Dimensions	5.5" L (14 cm)
17) Calibration Requirements/ Maintenance	NR
18) Manufacturer	Corning
19) Cost	$435 for electrode; $1,995 for meter
20) Comments	Double-junction reference and sensing half-cells; glass body; meter has automatic temperature compensation from −30 to 130°C

Solomat 601IS Fluoride Electrode with 520C Display

1) Analyte	Fluoride
2) Method	Ion-selective electrode
3) Range/Sensitivity	0.02 to 19,000 ppm
4) Accuracy	NR
5) Repeatability/Life Expectancy	NR
6) Zero Drift @ Constant Temp.	NR
7) Zero Drift @ Ambient Temp.	NR
8) Zero Noise @ Constant Temp.	NR
9) External Power Supply	110 VAC for display meter
10) Power Consumption	NR
11) Output Signal	mV
12) Electronic Response Time	NR
13) Operating Temperature Range	14 to 122°F (-10 to 50°C)
14) Storage Temperature Range/ Requirements	-5 to 140°F (-20 to 60°C)
15) Weight	1.1 lb for analyzer
16) External Dimensions	Analyzer is 2.9" W x 1.6" D increasing to 4.4" W x 2.6" D and is 8.3" L
17) Calibration Requirements/ Maintenance	NR
18) Manufacturer	Solomat
19) Cost	$629 for electrode; $195 for reference electrode; $1,395 for display meter
20) Comments	BNC plug

OMEGA Fluoride Electrode (based on use with Model PHI-359 meter)

1)	**Analyte**	Fluoride
2)	**Method**	Solid-state combination electrode; single-junction
3)	**Range/Sensitivity**	0.02 ppm to saturated for electrode; 0.001 to 1990 for meter; resolution to 0.001 for concentration and to 0.1 mV or 1 mV
4)	**Accuracy**	±0.5% for concentration; ±0.2% ±1 digit for mV
5)	**Repeatability/Life Expectancy**	NR
6)	**Zero Drift @ Constant Temp.**	NR
7)	**Zero Drift @ Ambient Temp.**	NR
8)	**Zero Noise @ Constant Temp.**	NR
9)	**External Power Supply**	9 V battery or 115 VAC adaptor
10)	**Power Consumption**	NR
11)	**Output Signal**	Recorder output is 0 to 200 mV
12)	**Electronic Response Time**	20 sec
13)	**Operating Temperature Range**	0 to 80°C for sensor; −30 to 130°C for meter
14)	**Storage Temperature Range/ Requirements**	NR
15)	**Weight**	2.9 kg (6.4 lb)
16)	**External Dimensions**	Meter is 211 L x 305 W x 104 mm H (8.3" x 12" x 4.1")
17)	**Calibration Requirements/ Maintenance**	Two-point autocalibration
18)	**Manufacturer**	OMEGA
19)	**Cost**	$186 for epoxy electrode and $275 for glass electrode; $1,260 for meter
20)	**Comments**	pH range is 5 to 8; KCl as reference electrolyte; interferences include OH^-; automatic or manual temperature compensation from 0 to 99.9°C

Cole-Parmer Fluoride

1) Analyte	Fluoride
2) Method	Solid-state electrode
3) Range/Sensitivity	0.1 to 260 ppm for electrode; meter is 0 to 9999 for concentration and ±1999 for mV
4) Accuracy	±0.5% of reading for concentration; ±0.2 for mV
5) Repeatability/Life Expectancy	NR
6) Zero Drift @ Constant Temp.	NR
7) Zero Drift @ Ambient Temp.	NR
8) Zero Noise @ Constant Temp.	NR
9) External Power Supply	110 VAC, 60 Hz; or 220 VAC, 50 Hz
10) Power Consumption	NR
11) Output Signal	±1999 mV, 1 mV/mV unit
12) Electronic Response Time	NR
13) Operating Temperature Range	0 to 80°C
14) Storage Temperature Range/ Requirements	NR
15) Weight	3.0 lb (1.4 kg) shipping weight
16) External Dimensions	9" L x 7" W x 2 3/8" H
17) Calibration Requirements/ Maintenance	NR
18) Manufacturer	Cole-Parmer
19) Cost	$290 for electrode; $795 for meter
20) Comments	pH range is 5 to 8; interferences are OH^- ; manual or automatic temperature compensation from −5 to 105°C

Capital Controls Aztec® F1000 Fluoride Monitor

1)	**Analyte**	Fluoride
2)	**Method**	Ion-selective electrode
3)	**Range/Sensitivity**	Automatic ranging from 0 to 10 mg/L as F^-; display resolution up to 0.001 mg/L
4)	**Accuracy**	2% of reading
5)	**Repeatability/Life Expectancy**	NR
6)	**Zero Drift @ Constant Temp.**	NR
7)	**Zero Drift @ Ambient Temp.**	NR
8)	**Zero Noise @ Constant Temp.**	NR
9)	**External Power Supply**	Automatic power recognition from 100 to 250 VAC, 47 to 63 Hz, 1 phase
10)	**Power Consumption**	480 W
11)	**Output Signal**	Single 4 to 20 mADC, 0 to 20 mADC, or 0 to 10 mADC isolated into 1000 ohms maximum
12)	**Electronic Response Time**	Speed of response is 90% step change within 5 min
13)	**Operating Temperature Range**	32 to 120°F (0 to 50°C) sample temperature; sample condition brings temperature to 35°C (96°F)
14)	**Storage Temperature Range/ Requirements**	32 to 140°F (0 to 60°C)
15)	**Weight**	35 lb (15 kg) shipping weight
16)	**External Dimensions**	21.5" x 26.5" x 8" (548 mm x 664 mm x 203 mm)
17)	**Calibration Requirements/ Maintenance**	Automatic high and low calibration with fluoride standard solutions; chemical reagents include buffering reagent, low- and high-fluoride standard solutions, and fluoride probe filling solution
18)	**Manufacturer**	Capital Controls Aztec®
19)	**Cost**	$18,000; $465/month for solutions
20)	**Comments**	Automatic ranging to 10 mg/L; self-diagnostics; alarm relays; sample flow of 200 to 500 mL/min; samples containing particles greater than 100 microns (0.004 inch) in diameter may require prefiltration

Wedgewood Model AF20 Series VIS/NIR In-Line Sensor and Model 680 VIS/NIR Analyzer Transmitter

1)	**Analyte**	Color
2)	**Method**	Spectral absorbence; enhanced silicon/germanium detectors
3)	**Range/Sensitivity**	Five fixed selectable ranges or one variable range; 400 to 1,900 nm
4)	**Accuracy**	2% over full range
5)	**Repeatability/Life Expectancy**	0.5% (any range)
6)	**Zero Drift @ Constant Temp.**	NR
7)	**Zero Drift @ Ambient Temp.**	NR
8)	**Zero Noise @ Constant Temp.**	NR
9)	**External Power Supply**	115/230 VAC ±15%, 50/60 Hz, 12 W
10)	**Power Consumption**	NR
11)	**Output Signal**	Two simultaneous, 0 to 10 VDC, 0 to 20/4 to 20 mA
12)	**Electronic Response Time**	
13)	**Operating Temperature Range**	0 to 100°C for sensor; 0 to 70°C for analyzer
14)	**Storage Temperature Range/ Requirements**	-20 to 70°C for analyzer
15)	**Weight**	Analyzer is less than 3 kg
16)	**External Dimensions**	Analyzer is 5.6 x 5.06 x 6.69"
17)	**Calibration Requirements/ Maintenance**	Incandescent light with 10,000- to 25,000-hr lifespan
18)	**Manufacturer**	Wedgewood Technology, Inc.
19)	**Cost**	$6,000 to $7,000
20)	**Comments**	Output in optical density units; optical filters selected based on spectral range of sample; alarms; temperature compensation from 0 to 25°C (0 to 1% per °C); two selected wavelengths from 400 nm to 1,900 nm; stainless steel sensor

Tytronics® Sentinel Color Analyzer

1) Analyte	Apparent color
2) Method	Direct spectrophotometric
3) Range/Sensitivity	450 nm and 810 nm
4) Accuracy	NR
5) Repeatability/Life Expectancy	NR
6) Zero Drift @ Constant Temp.	NR
7) Zero Drift @ Ambient Temp.	NR
8) Zero Noise @ Constant Temp.	NR
9) External Power Supply	NR
10) Power Consumption	NR
11) Output Signal	4 to 20 mA
12) Electronic Response Time	NR
13) Operating Temperature Range	NR
14) Storage Temperature Range/ Requirements	NR
15) Weight	NR
16) External Dimensions	NR
17) Calibration Requirements/ Maintenance	Automatic calibration; recommended weekly
18) Manufacturer	Tytronics®
19) Cost	$8,950 for single stream
20) Comments	Batch-sampling technique; alarm functions; designed for potable water applications

MONITEK® CS6 (with MONITEK® CT9 Transmitter)

1) Analyte	Color
2) Method	Sensor with in-line construction detects incandescent light; sensor measures absorbence in a liquid stream by using two side-by-side silicon photodetectors; instrument is actually a monitor that measures color differences; however, if calibrated with known highs and lows, it can be set up to give absolute readings
3) Range/Sensitivity	Full-scale ranges for transmitter are 0 to 100%, 0.01 to 5.0 absorbence units (A.U.), and other ranges available; sensor sensitivity is ±0.1% of span
4) Accuracy	Resolution is 0.02% of full scale; linearity is 0.1%
5) Repeatability/Life Expectancy	0.1% of full scale
6) Zero Drift @ Constant Temp.	NR
7) Zero Drift @ Ambient Temp.	NR
8) Zero Noise @ Constant Temp.	NR
9) External Power Supply	90 to 265 VAC, 50 to 60 Hz, 35 W
10) Power Consumption	NR
11) Output Signal	4 to 20 mA; 0 to 20 mA; or 0 to 1 mA
12) Electronic Response Time	Can have online real-time instantaneous reading
13) Operating Temperature Range	Process temperature of 0 to 250°C for sensor 0 to 50°C
14) Storage Temperature Range/ Requirements	0 to 50°C ambient temperature for transmitter
15) Weight	Sensor is 5 lb (2.3 kg); transmitter is 25 lb (11.4 kg)
16) External Dimensions	Sensor mountings available for 0.5" or 1" NPT; transmitter is 14.7" x 12.3" x 6.7"
17) Calibration Requirements/ Maintenance	Can use coating to prevent scaling on window depending on application; change incandescent light bulb as needed
18) Manufacturer	MONITEK®
19) Cost	~$9,000 (low-range CS8 sensor costs ~$12,000)
20) Comments	Detector is silicon photovoltaic cell; light source is 400 to 1100 nm incandescent lamp: samples can be sent to their lab at no charge to determine if low-range sensor is required; line size ranging from 0.5" to 6" can be selected; standard with four relays; optional communication ports to run output directly to PC

Component Engineering

1) Analyte	Color
2) Method	Potentiometers
3) Range/Sensitivity	Color sensors are available for various target distances and sizes; spectral sensitivity of 400 to 700 nm
4) Accuracy	Depends on what is being monitored (i.e., if it is liquid, moving, or contains air bubbles)
5) Repeatability/Life Expectancy	NR
6) Zero Drift @ Constant Temp.	NR
7) Zero Drift @ Ambient Temp.	NR
8) Zero Noise @ Constant Temp.	NR
9) External Power Supply	Supply voltage is 20 to 30 VDC; maximum load current is 50 mA; supply for the type FW 01 wenglor white light source is provided by a 6V/2A outlet on the FN01 color analyzer unit
10) Power Consumption	NR
11) Output Signal	200 mA maximum
12) Electronic Response Time	0.5 ms
13) Operating Temperature Range	−10 to 60°C
14) Storage Temperature Range/ Requirements	NR
15) Weight	NR
16) External Dimensions	Sensor has 18 mm diameter; can be reduced to 6 to 8 mm with fiber optics on end
17) Calibration Requirements/ Maintenance	NR
18) Manufacturer	Component Engineering
19) Cost	~$1,900 for light source and sensor with fiber optics; may be able to use sensor without fiber optics which would reduce price by ~$200
20) Comments	Optical fiber components

Solomat 635IS Water Hardness Electrode with 520C Display

1) Analyte	Water hardness
2) Method	Ion-selective electrode
3) Range/Sensitivity	0.4 to 40,000 ppm
4) Accuracy	NR
5) Repeatability/Life Expectancy	NR
6) Zero Drift @ Constant Temp.	NR
7) Zero Drift @ Ambient Temp.	NR
8) Zero Noise @ Constant Temp.	NR
9) External Power Supply	110 VAC for display meter
10) Power Consumption	NR
11) Output Signal	mV
12) Electronic Response Time	NR
13) Operating Temperature Range	14 to 122°F (−10 to 50°C)
14) Storage Temperature Range/ Requirements	−5 to 140°F (−20 to 60°C)
15) Weight	1.1 lb for analyzer
16) External Dimensions	Analyzer is 2.9" W x 1.6" D, increasing to 4.4" W x 2.6" D, and is 8.3" L
17) Calibration Requirements/ Maintenance	NR
18) Manufacturer	Solomat
19) Cost	\$829 for electrode; \$195 for reference electrode; \$1,395 for display meter
20) Comments	BNC plug

Orion Model 1820

1)	**Analyte**	Hardness
2)	**Method**	Calcium ISE and patented passive diffusion reagent addition system
3)	**Range/Sensitivity**	Up to 350 ppm calcium hardness as $CaCO_3$
4)	**Accuracy**	±10% of reading or ±10 ppb as $CaCO_3$, whichever is greater, within ±10°C of calibration equipment
5)	**Repeatability/Life Expectancy**	Precision is ±5% of reading or ±5 ppb, whichever is greater
6)	**Zero Drift @ Constant Temp.**	10 ppb or 10% of reading per month, whichever is greater
7)	**Zero Drift @ Ambient Temp.**	NR
8)	**Zero Noise @ Constant Temp.**	±1 ppb or 1% of reading per month, whichever is greater
9)	**External Power Supply**	100/115/220 VAC ±10%, 50/60 Hz,100 W
10)	**Power Consumption**	NR
11)	**Output Signal**	NR
12)	**Electronic Response Time**	50% within 2 min, function of flow and volume
13)	**Operating Temperature Range**	5 to 40°C (41 to 104°F)
14)	**Storage Temperature Range/ Requirements**	5 to 45°C (41 to 113°F); 95% relative humidity, noncondensing
15)	**Weight**	20.9 kg (46 lb)
16)	**External Dimensions**	Analyzer is 65 x 45 x 27 cm (26 x 17 x 11")
17)	**Calibration Requirements/ Maintenance**	Requires 20 min of maintenance per month; calibration with patented double known addition (DKA) technique; minimum calibration once per month; electrode life of 6 months minimum; 1 L reagent use every 2 months
18)	**Manufacturer**	Orion
19)	**Cost**	$4,245 for analyzer; $1,050 for consumables kit (1-yr supply); $575 for calibration kit (1 time)
20)	**Comments**	Optional alarm module

Hach Model 1220

1) Analyte	Hardness
2) Method	Buffered wet-chemical colorimetric indicator to detect total hardness
3) Range/Sensitivity	0 to 100 mg/L
4) Accuracy	NR
5) Repeatability/Life Expectancy	±10% of alarm trip value on 1 and 2 mg/L range; 4% of alarm trip value on remaining ranges
6) Zero Drift @ Constant Temp.	NR
7) Zero Drift @ Ambient Temp.	NR
8) Zero Noise @ Constant Temp.	NR
9) External Power Supply	115/230 VAC, 50/60 Hz, selectable, 18 W
10) Power Consumption	NR
11) Output Signal	NR
12) Electronic Response Time	NR
13) Operating Temperature Range	0 to 40°C
14) Storage Temperature Range/ Requirements	NR
15) Weight	14 kg (32 lb) shipping weight
16) External Dimensions	35.6 x 30.5 x 15.8 cm (14 x 12 x 6.25"); wall mounting
17) Calibration Requirements/ Maintenance	Requires 1 L each of indicator and buffer which need to be replenished every 2 months for continuous operation
18) Manufacturer	Hach
19) Cost	$1,500
20) Comments	Alarm system — can select from trip point of 1, 2, 5, 10, 20, 50, or 100 mg/L

Cole-Parmer Water Hardness

1)	**Analyte**	Hardness
2)	**Method**	PVC membrane combination electrode
3)	**Range/Sensitivity**	0.4 to 40,000 ppm as Ca^{2+} for electrode; meter is 0 to 9999 for concentration and ±1999 for mV
4)	**Accuracy**	±0.5% of reading for concentration; ±0.2 for mV
5)	**Repeatability/Life Expectancy**	NR
6)	**Zero Drift @ Constant Temp.**	NR
7)	**Zero Drift @ Ambient Temp.**	NR
8)	**Zero Noise @ Constant Temp.**	NR
9)	**External Power Supply**	110 VAC, 60 Hz; or 220 VAC, 50 Hz
10)	**Power Consumption**	NR
11)	**Output Signal**	±1999 mV, 1 mV/mV unit
12)	**Electronic Response Time**	NR
13)	**Operating Temperature Range**	0 to 50°C
14)	**Storage Temperature Range/ Requirements**	NR
15)	**Weight**	3.0 lb (1.4 kg) shipping weight
16)	**External Dimensions**	9" L x 7" W x 2^3/8" H
17)	**Calibration Requirements/ Maintenance**	NR
18)	**Manufacturer**	Cole-Parmer
19)	**Cost**	$340 for electrode; $795 for meter
20)	**Comments**	pH range is 5 to 10; interferences are Cu^{+2}, Zn^{+2}, Ni^{+2}, and $Fe^{+2;}$ single- or double-junction electrodes; manual or automatic temperature compensation from −5 to 105°C

OMEGA Water Hardness (based on use with Model PHI-359 meter)

1)	**Analyte**	Hardness
2)	**Method**	Polymer membrane combination electrode; single-junction
3)	**Range/Sensitivity**	0.4 to 40,000 ppm for electrode; 0.001 to 1990 for meter; resolution to 0.001 for concentration and to 0.1 mV or 1 mV
4)	**Accuracy**	±0.5% for concentration; ±0.2% ±1 digit for mV
5)	**Repeatability/Life Expectancy**	NR
6)	**Zero Drift @ Constant Temp.**	NR
7)	**Zero Drift @ Ambient Temp.**	NR
8)	**Zero Noise @ Constant Temp.**	NR
9)	**External Power Supply**	9 V battery or 115 VAC adaptor
10)	**Power Consumption**	NR
11)	**Output Signal**	Recorder output is 0 to 200 mV
12)	**Electronic Response Time**	30 sec
13)	**Operating Temperature Range**	0 to 50°C for sensor; −30 to 130°C for meter
14)	**Storage Temperature Range/ Requirements**	NR
15)	**Weight**	2.9 kg (6.4 lb)
16)	**External Dimensions**	Meter is 211 L x 305 W x 104 mm H (8.3" x 12" x 4.1")
17)	**Calibration Requirements/ Maintenance**	Two-point autocalibration
18)	**Manufacturer**	OMEGA
19)	**Cost**	$220 for PVC electrode and $275 for glass electrode; $1,260 for meter
20)	**Comments**	pH range is 5 to 10; KCl as reference electrolyte; interferences include copper, zinc, nickel, and iron; automatic or manual temperature compensation from 0 to 99.9°C

Hach 6000 Low Range Hardness Analyzer

1)	**Analyte**	Hardness
2)	**Method**	Total Hardness Method 130.1 (colorimetric, automated EDTA)
3)	**Range/Sensitivity**	50 to 10,000 μg/L
4)	**Accuracy**	NR
5)	**Repeatability/Life Expectancy**	NR
6)	**Zero Drift @ Constant Temp.**	NR
7)	**Zero Drift @ Ambient Temp.**	NR
8)	**Zero Noise @ Constant Temp.**	NR
9)	**External Power Supply**	120 or 240 VAC
10)	**Power Consumption**	NR
11)	**Output Signal**	4 to 20 mA
12)	**Electronic Response Time**	Less than 5-min cycle time
13)	**Operating Temperature Range**	5 to 50°C sample temperature range
14)	**Storage Temperature Range/ Requirements**	NR
15)	**Weight**	NR
16)	**External Dimensions**	NR
17)	**Calibration Requirements/ Maintenance**	Automatic calibration
18)	**Manufacturer**	Hach
19)	**Cost**	$5,800
20)	**Comments**	Critical components are automatically primed, cleaned, and calibrated; programmable alarms

Orion 142

1)	**Analyte**	Conductivity, TDS, salinity, and temperature
2)	**Method**	NR
3)	**Range/Sensitivity**	Conductivity range is 0 to 500,000 μs/cm and resolution is 0.01 μs/cm; TDS range is 0 to 1999 mg/L and resolution is 1 mg/L
4)	**Accuracy**	Conductivity is 0.5% ±1 digit at 15 to 35°C; TDS is ±2% ±1 digit at 15 to 35°C
5)	**Repeatability/Life Expectancy**	NR
6)	**Zero Drift @ Constant Temp.**	NR
7)	**Zero Drift @ Ambient Temp.**	NR
8)	**Zero Noise @ Constant Temp.**	NR
9)	**External Power Supply**	Rechargeable battery, kAC line, 110, 220, or 240 V
10)	**Power Consumption**	NR
11)	**Output Signal**	70 mA, RS232 interface
12)	**Electronic Response Time**	NR
13)	**Operating Temperature Range**	−10 to 55°
14)	**Storage Temperature Range/ Requirements**	NR
15)	**Weight**	1.5 kg
16)	**External Dimensions**	200 mm x 90 mm x 190 mm
17)	**Calibration Requirements/ Maintenance**	Automated calibration
18)	**Manufacturer**	Orion
19)	**Cost**	$1,650
20)	**Comments**	Four temperature compensation algorithms

Honeywell

1) **Analyte**	Turbidity, conductivity, temperature
2) **Method**	Conductivity is measured using two tin-over-nickel-plated 303 stainless steel probes emitting a low-amplitude AC signal
3) **Range/Sensitivity**	0.00001 to 15 millisiemens (mS) for conductivity; 0 to 4000 NTU for turbidity; 68 to 140°F for temperature
4) **Accuracy**	See manufacturer's curve for turbidity and conductivity
5) **Repeatability/Life Expectancy**	NR
6) **Zero Drift @ Constant Temp.**	NR
7) **Zero Drift @ Ambient Temp.**	NR
8) **Zero Noise @ Constant Temp.**	NR
9) **External Power Supply**	8 to 30 VDC power supply
10) **Power Consumption**	NR
11) **Output Signal**	Data transmitted via a bidirectional 5-VDC RS232 serial communications link
12) **Electronic Response Time**	1.3 sec for turbidity; 0.03 sec response time and 3 to 5 min stabilization time for temperature; 0.85 sec for conductivity
13) **Operating Temperature Range**	68 to 140°F; may be wider range, but has been tested only within this range
14) **Storage Temperature Range/ Requirements**	NR
15) **Weight**	NR
16) **External Dimensions**	Mounting dimensions are 1.25" diameter x 2.39"; cap is 0.71" OD x 2.75"
17) **Calibration Requirements/ Maintenance**	NR
18) **Manufacturer**	Honeywell
19) **Cost**	$90 without flowthrough cap; $100 with flowthrough cap
20) **Comments**	With microprocessor and software

Hanna Instruments HI 9635

1) Analyte	Conductivity, TDS, and temperature
2) Method	NR
3) Range/Sensitivity	0.0 to 150.0 ppm (mg/L), 150 to 1500 ppm (mg/L), 1.50 to 15.00 ppt (g/L); resolutions are 0.1 ppm, 1 ppm, and 0.01 ppt
4) Accuracy	±2% full scale excluding probe error (@20°C)
5) Repeatability/Life Expectancy	Typical deviation ±2% full scale
6) Zero Drift @ Constant Temp.	NR
7) Zero Drift @ Ambient Temp.	NR
8) Zero Noise @ Constant Temp.	NR
9) External Power Supply	Power socket for 12-VDC or 4 x 1.5-V AAA batteries
10) Power Consumption	NR
11) Output Signal	NR
12) Electronic Response Time	NR
13) Operating Temperature Range	32 to 122°F (0 to 50°C); relative humidity (RH) 100%
14) Storage Temperature Range/ Requirements	NR
15) Weight	15 oz (425 g)
16) External Dimensions	7.7 x 3.1 x 2.4" (196 x 80 x 60 mm); probe has 1-m screened cable
17) Calibration Requirements/ Maintenance	Automatic 1-, 2-, or 3-point calibration at 84, 1413, and 12,880 μs/cm
18) Manufacturer	Hanna Instruments, Inc. (HI)
19) Cost	$490
20) Comments	PVC probe, automatic temperature compensation from 32 to 140°F (0 to 60°C) with an adjustable β from 0.0 to 3.0% per degree °C and TDS factor at 0.70 or 0.50

Orion 124 Hand-Held Conductivity/TDS Meter

1)	**Analyte**	Conductivity, TDS, temperature
2)	**Method**	Four-electrode cell technology
3)	**Range/Sensitivity**	0 to 2,000 ppm TDS; 0 to 199,000 μS/cm in four ranges; −5 to 50°C temperature
4)	**Accuracy**	±0.5% for conductivity; ±0.1°C for temperature
5)	**Repeatability/Life Expectancy**	NR
6)	**Zero Drift @ Constant Temp.**	NR
7)	**Zero Drift @ Ambient Temp.**	NR
8)	**Zero Noise @ Constant Temp.**	NR
9)	**External Power Supply**	9-VDC alkaline battery for meter
10)	**Power Consumption**	NR
11)	**Output Signal**	NR
12)	**Electronic Response Time**	NR
13)	**Operating Temperature Range**	NR
14)	**Storage Temperature Range/ Requirements**	NR
15)	**Weight**	NR
16)	**External Dimensions**	17 x 9 x 6 cm for meter; 21 mm diameter x 150 mm long for cell
17)	**Calibration Requirements/ Maintenance**	NR
18)	**Manufacturer**	Orion
19)	**Cost**	$795 for meter; $260 for cell
20)	**Comments**	Automatic temperature compensation; solids factor adjustable from 0.5 to 1.0

OMEGA CDH-287-KIT

1)	**Analyte**	Conductivity, TDS, resistivity, salinity
2)	**Method**	NR
3)	**Range/Sensitivity**	0 to 19.99 mg/L, 0 to 199.9 mg/L, 0 to 1999 mg/L, 0 to 19.99 g/L, 0 to 130 g/L for TDS; 0 to 19.99 μS/cm; 0 to 199.9 μS/cm, 0 to 1999 μS/cm, 0 to 19.99 mS/cm, 0 to 199.9 mS/cm, 0 to 1999 mS/cm for conductivity
4)	**Accuracy**	±0.5% of reading
5)	**Repeatability/Life Expectancy**	NR
6)	**Zero Drift @ Constant Temp.**	NR
7)	**Zero Drift @ Ambient Temp.**	NR
8)	**Zero Noise @ Constant Temp.**	NR
9)	**External Power Supply**	9-V battery
10)	**Power Consumption**	NR
11)	**Output Signal**	±200.0 mV
12)	**Electronic Response Time**	NR
13)	**Operating Temperature Range**	0 to 50°C
14)	**Storage Temperature Range/ Requirements**	NR
15)	**Weight**	410 g (0.9 lb)
16)	**External Dimensions**	100 x 180 x 44 mm
17)	**Calibration Requirements/ Maintenance**	NR
18)	**Manufacturer**	OMEGA
19)	**Cost**	Meter with probe, $740; replacement probes, $105 to $275
20)	**Comments**	Automatic temperature compensation from 0 to 50°C

Orion Advanced Portable Multimeter

1)	**Analyte**	Conductivity, salinity, TDS, temperature
2)	**Method**	NR
3)	**Range/Sensitivity**	0 to 1,999,900 μS in six ranges; −5 to 90°C for temperature
4)	**Accuracy**	0.5% ±1 digit for conductivity; ±0.5°C for temperature
5)	**Repeatability/Life Expectancy**	Resolution 0.01 μS
6)	**Zero Drift @ Constant Temp.**	NR
7)	**Zero Drift @ Ambient Temp.**	NR
8)	**Zero Noise @ Constant Temp.**	NR
9)	**External Power Supply**	120-VAC line converter
10)	**Power Consumption**	NR
11)	**Output Signal**	NR
12)	**Electronic Response Time**	NR
13)	**Operating Temperature Range**	NR
14)	**Storage Temperature Range/ Requirements**	NR
15)	**Weight**	NR
16)	**External Dimensions**	NR
17)	**Calibration Requirements/ Maintenance**	NR
18)	**Manufacturer**	Orion
19)	**Cost**	$1,650 for meter; $330 for standard cell
20)	**Comments**	Autoranging; automatic temperature compensation

Davis Instruments Model 72

1) Analyte	Conductivity, TDS, temperature
2) Method	Carbon electrode
3) Range/Sensitivity	10, 100, 1,000, and 10,000 μS/cm for conductivity; 7, 70, 700, and 7,000 μg/L for TDS; −50 to 150°C for temperature
4) Accuracy	±2.5%
5) Repeatability/Life Expectancy	NR
6) Zero Drift @ Constant Temp.	NR
7) Zero Drift @ Ambient Temp.	NR
8) Zero Noise @ Constant Temp.	NR
9) External Power Supply	9-V battery; optional AC jack
10) Power Consumption	NR
11) Output Signal	NR
12) Electronic Response Time	NR
13) Operating Temperature Range	NR
14) Storage Temperature Range/ Requirements	NR
15) Weight	4 lb
16) External Dimensions	Cell is 6" L x 0.5" diameter
17) Calibration Requirements/ Maintenance	NR
18) Manufacturer	Davis Instruments
19) Cost	$360 for meter; $94 for conductivity cell
20) Comments	Manual temperature compensation from 0 to 100°C

Hanna Instruments HI 8734

1)	**Analyte**	TDS
2)	**Method**	NR
3)	**Range/Sensitivity**	0 to 19.99 g/L over 3 ranges; 0 to 199.9 mg/L, 0 to 1999 mg/L, 0 to 19.99 g/L
4)	**Accuracy**	±1% full scale
5)	**Repeatability/Life Expectancy**	NR
6)	**Zero Drift @ Constant Temp.**	NR
7)	**Zero Drift @ Ambient Temp.**	NR
8)	**Zero Noise @ Constant Temp.**	NR
9)	**External Power Supply**	9-V battery
10)	**Power Consumption**	NR
11)	**Output Signal**	NR
12)	**Electronic Response Time**	NR
13)	**Operating Temperature Range**	NR
14)	**Storage Temperature Range/ Requirements**	NR
15)	**Weight**	370 g for meter and probe
16)	**External Dimensions**	180 L x 83 W x 40 H mm
17)	**Calibration Requirements/ Maintenance**	NR
18)	**Manufacturer**	Hanna Instruments, Inc.
19)	**Cost**	$298 for meter and probe; $65 for replacement probe
20)	**Comments**	Manual temperature compensation from 0 to 50°C

Cole-Parmer Conductivity/TDS Meter

1) Analyte	Conductivity, TDS
2) Method	SS electrode
3) Range/Sensitivity	0.00 to 19.99 µS, 0.0 to 199.9 µS, 0 to 1999 µS, 0.00 to 19.99 mS for conductivity; 0.00 to 9.99 ppm, 0.0 to 99.9 ppm, 0 to 999 ppm, 1.00 to 9.99 ppt for TDS; 0 to 80.0°C for temperature
4) Accuracy	Greater of ±1% or ±1 digit for conductivity and TDS; ±0.5°C for temperature
5) Repeatability/Life Expectancy	NR
6) Zero Drift @ Constant Temp.	NR
7) Zero Drift @ Ambient Temp.	NR
8) Zero Noise @ Constant Temp.	NR
9) External Power Supply	Four 1.5-V batteries
10) Power Consumption	NR
11) Output Signal	NR
12) Electronic Response Time	NR
13) Operating Temperature Range	0 to 50°C
14) Storage Temperature Range/ Requirements	NR
15) Weight	1.3 lb (0.6 kg) shipping weight
16) External Dimensions	110 mm L x 12 mm OD for cell; 3.5" W x 7.375" H x 1.75" D
17) Calibration Requirements/ Maintenance	NR
18) Manufacturer	Cole-Parmer
19) Cost	$395 for meter; $65 for replacement cell
20) Comments	Automatic ranging in four ranges; automatic temperature compensation from 0 to 50°C

Hach Model 1900 WPC Process Particle Counter

1) Analyte	Particle count
2) Method	Light-extinction technology to size and count particles; optical system has infrared laser diode
3) Range/Sensitivity	Detects particles from 2 to 800 microns in size, with particle concentration ranges of 0 to 20,000 particles/mL
4) Accuracy	NR
5) Repeatability/Life Expectancy	NR
6) Zero Drift @ Constant Temp.	NR
7) Zero Drift @ Ambient Temp.	NR
8) Zero Noise @ Constant Temp.	NR
9) External Power Supply	95 to 240 VAC, 50/60 Hz
10) Power Consumption	NR
11) Output Signal	Five 4- to 20-mA signals representing particle counts in size ranges, total count, and flow or sensor status; can optionally accept two 4- to 20-mA output signals from other instruments
12) Electronic Response Time	NR
13) Operating Temperature Range	Sample temperature of 0 to 50°C (32 to 122°F)
14) Storage Temp. Range/ Requirements	0 to 40°C (32 to 104°F)
15) Weight	2.3 kg (5 lb)
16) External Dimensions	14 x 23 x 10.2 cm (5.5 x 9 x 4")
17) Calibration Requirements/ Maintenance	NR
18) Manufacturer	Hach
19) Cost	$9,990 (two-sensor system); $6,325 (single-sensor system)
20) Comments	Sensors placed at filter bed inlet and outlet; software program displays sample information and log reduction or percent filter efficiency; alarm capabilities; 14" of head pressure required

Met One PCX

1) Analyte	Particle count
2) Method	Light blocking; laser diode light source; electronic photodiode
3) Range/Sensitivity	Up to 32 particle sizes from 2 microns up to 750 microns
4) Accuracy	NR
5) Repeatability/Life Expectancy	NR
6) Zero Drift @ Constant Temp.	NR
7) Zero Drift @ Ambient Temp.	NR
8) Zero Noise @ Constant Temp.	NR
9) External Power Supply	115 VAC, 50/60 Hz; optional 230 VAC, 50/60 Hz
10) Power Consumption	NR
11) Output Signal	4 to 20 mA, 0 to 5 / 0 to 10 V
12) Electronic Response Time	NR
13) Operating Temperature Range	32 to 122°F (0 to 50°C) for sensor; 32 to 140°F (0 to 60°C) for power supply
14) Storage Temp. Range/ Requirements	NR
15) Weight	7.4 lb for sensor; 3.8 lb for power supply
16) External Dimensions	11.0" H x 9.0" W x 6.7" D for sensor; 7.0" H x 7.0" W x 6.7" D for power supply
17) Calibration Requirements/ Maintenance	NR
18) Manufacturer	Met One
19) Cost	$4,100 with flow controller ($3,950 without)
20) Comments	100 mL/min recommended sample flowrate

C

COTS Devices for System Operation Monitoring and Control

The COTS devices for system operation monitoring and control include the various sensors, pumps, and valves referenced in the tables presented in Chapter 5.

This listing is provided solely for informational purposes. Inclusion in this appendix does not constitute recommendation of the companies listed or endorsement for use of their products. Reasonable efforts have been made to ensure that the information provided is up to date.

Flowmetrics Series 2000

1) Analyte	Flow monitor
2) Method	Mechanical rotameter
3) Range/Sensitivity	180 gpm maximum capacity for 2" pipe
4) Accuracy	±2% full scale
5) Repeatability/Life Expectancy	Repeatability of 0.5% full scale
6) Zero Drift @ Constant Temp.	NR
7) Zero Drift @ Ambient Temp.	NR
8) Zero Noise @ Constant Temp.	NR
9) External Power Supply	None needed
10) Power Consumption	Does not apply
11) Output Signal	Does not apply
12) Electronic Response Time	Does not apply
13) Operating Temperature Range	250°F maximum operating temperature
14) Storage Temperature Range/ Requirements	NR
15) Weight	26.5 lb
16) External Dimensions	24" L x 4.75" W x 4.75" D
17) Calibration Requirements/ Maintenance	Factory calibration
18) Manufacturer	Flowmetrics
19) Cost	Series 2000 with 2" tube size, ~$850
20) Comments	Visual rotameter readings only; large, bulky device; stainless steel construction for durability; not easily configured to electronics for remote monitoring

OMEGA FLR8313-I

1)	**Analyte**	Flow monitor/totalizer
2)	**Method**	Rotameter with electronic output
3)	**Range/Sensitivity**	10 to 130 gpm for 1.25" pipe
4)	**Accuracy**	2 to 5% FS
5)	**Repeatability/Life Expectancy**	NR
6)	**Zero Drift @ Constant Temp.**	NR
7)	**Zero Drift @ Ambient Temp.**	NR
8)	**Zero Noise @ Constant Temp.**	NR
9)	**External Power Supply**	120 VAC
10)	**Power Consumption**	NR
11)	**Output Signal**	4 to 20 mA analog output
12)	**Electronic Response Time**	Continuous output
13)	**Operating Temperature Range**	210°F maximum operating temperature
14)	**Storage Temperature Range/ Requirements**	NR
15)	**Weight**	NR
16)	**External Dimensions**	12.2" L x 4" W x 5.4" H
17)	**Calibration Requirements/ Maintenance**	Factory calibration; on-site zero and span adjustments
18)	**Manufacturer**	OMEGA
19)	**Cost**	Model FLR8313-I, $1,145
20)	**Comments**	Rotameter design, although simple, may be prone to fouling; limited by the pipe size available to ~1.25"; flowrates of a treatment plant may exceed upper limits of a rotameter design; additional electronics required for a totalizer

MC Meters and Controls Company

1)	**Analyte**	Flow/totalizer
2)	**Method**	Turbine meter with cast-iron body
3)	**Range/Sensitivity**	4 to 350 gpm for 2" pipe
4)	**Accuracy**	±1.5%
5)	**Repeatability/Life Expectancy**	NR
6)	**Zero Drift @ Constant Temp.**	NR
7)	**Zero Drift @ Ambient Temp.**	NR
8)	**Zero Noise @ Constant Temp.**	NR
9)	**External Power Supply**	NR
10)	**Power Consumption**	NR
11)	**Output Signal**	4 to 20 mA
12)	**Electronic Response Time**	Continuous sensing while flow exists
13)	**Operating Temperature Range**	NR
14)	**Storage Temperature Range/ Requirements**	NR
15)	**Weight**	24 lb
16)	**External Dimensions**	Laying length, 10" x 8" H x 6.25" W
17)	**Calibration Requirements/ Maintenance**	Factory calibration
18)	**Manufacturer**	MC Meters and Controls Company
19)	**Cost**	2" cast-iron turbine, $391; flowrate output register, $207; strainer, $225
20)	**Comments**	Flanged ends for incorporation into system piping; low head loss; 2-yr guarantee to perform to American Water Works Association (AWWA) new meter accuracy standard

Fluidyne Model 1100

1)	**Analyte**	Flow monitor/totalizer
2)	**Method**	Vortex-shedding flowmeter
3)	**Range/Sensitivity**	2.5 to 125 gpm for 2" pipe in a PVC tee
4)	**Accuracy**	±0.5% FS combined linearity and repeatability
5)	**Repeatability/Life Expectancy**	NR
6)	**Zero Drift @ Constant Temp.**	NR
7)	**Zero Drift @ Ambient Temp.**	NR
8)	**Zero Noise @ Constant Temp.**	NR
9)	**External Power Supply**	10 to 32 VDC
10)	**Power Consumption**	NR
11)	**Output Signal**	4 to 20 mA analog output
12)	**Electronic Response Time**	NR
13)	**Operating Temperature Range**	32° to 140°F process temperature
14)	**Storage Temperature Range/ Requirements**	-20° to 140°F ambient temperature
15)	**Weight**	NR
16)	**External Dimensions**	Tee type, 5.12" L x ~7" H; Flange type, 8.47" L x ~9" H
17)	**Calibration Requirements/ Maintenance**	Factory calibration; on-site zero and span adjustments made according to specification sheets and outputs
18)	**Manufacturer**	Fluidyne
19)	**Cost**	Model 1100 (in PVC tee) with local rate/total output, $850
20)	**Comments**	No moving parts; low flow capacity; rugged mechanical design ensures low maintenance cost; 2-yr warranty

EMCO Vortex PhD™

1)	**Analyte**	Flow/totalizer
2)	**Method**	Vortex-shedding flowmeter
3)	**Range/Sensitivity**	1.9 to 32 ft/sec velocities; 13.8 to 295 gpm for 2" pipe
4)	**Accuracy**	±0.7% of flowrate
5)	**Repeatability/Life Expectancy**	Repeatability of ±0.15% of flowrate
6)	**Zero Drift @ Constant Temp.**	NR
7)	**Zero Drift @ Ambient Temp.**	NR
8)	**Zero Noise @ Constant Temp.**	NR
9)	**External Power Supply**	120 VAC for flowrate indicators and totalizer
10)	**Power Consumption**	<1 W
11)	**Output Signal**	4 to 20 mA analog output, 2-wire
12)	**Electronic Response Time**	Adjustable from 1 to 100 sec (analog output version)
13)	**Operating Temperature Range**	-40 to 750°F process temperature limit
14)	**Storage Temperature Range/ Requirements**	-20 to 140°F ambient temp limits (electrical specs)
15)	**Weight**	NR
16)	**External Dimensions**	NR
17)	**Calibration Requirements/ Maintenance**	Factory calibration; on-site zero and span adjustments
18)	**Manufacturer**	EMCO
19)	**Cost**	Vortex PhD™ for 2" pipe, $1,471; local flowrate/totalizer, $150; miscellaneous installation parts, $375
20)	**Comments**	Flanged ends for incorporation into system piping; no moving parts to foul; corrosion-resistant, stainless steel construction; flow and totalizer pulses obtainable

OMEGA Series FMG 2550

1)	**Analyte**	Flow monitor
2)	**Method**	Magnetic flowmeter, insertion-type
3)	**Range/Sensitivity**	0.3 to 20 ft/sec flow velocity range for 2" to 12" pipe
4)	**Accuracy**	±2% of reading
5)	**Repeatability/Life Expectancy**	NR
6)	**Zero Drift @ Constant Temp.**	NR
7)	**Zero Drift @ Ambient Temp.**	NR
8)	**Zero Noise @ Constant Temp.**	NR
9)	**External Power Supply**	24 VDC; ±10%
10)	**Power Consumption**	NR
11)	**Output Signal**	Isolated 4 to 20 mA or 0 to 20 mA into 600-ohm maximum load
12)	**Electronic Response Time**	Continuous
13)	**Operating Temperature Range**	32 to 212°F fluid temperature
14)	**Storage Temperature Range/ Requirements**	-4 to 176°F
15)	**Weight**	11 lb
16)	**External Dimensions**	8.6" D x 12.9" H
17)	**Calibration Requirements/ Maintenance**	Factory calibration; 100% keypad calibration
18)	**Manufacturer**	OMEGA
19)	**Cost**	Series FMG 2550 Magmeter, $2,675; FPM-9010 electronics, $610
20)	**Comments**	Insertion-type magnetic flowmeter; requires minimum fluid conductivity of 5 μS/cm; easy installation into pipes from 2" to 12" diameter via 2" NPT fitting

OMEGA Series FMG 94

1)	**Analyte**	Flow monitor
2)	**Method**	Magnetic flowmeter, water style
3)	**Range/Sensitivity**	30 to 300 gpm for 2" pipe
4)	**Accuracy**	±1% of reading
5)	**Repeatability/Life Expectancy**	±0.25% of the rate
6)	**Zero Drift @ Constant Temp.**	NR
7)	**Zero Drift @ Ambient Temp.**	NR
8)	**Zero Noise @ Constant Temp.**	NR
9)	**External Power Supply**	115 VAC or 10 to 30 VDC
10)	**Power Consumption**	15 W maximum
11)	**Output Signal**	4 to 20 or 0 to 20 mADC analog range
12)	**Electronic Response Time**	Continuous
13)	**Operating Temperature Range**	0 to 200°F fluid temperature
14)	**Storage Temperature Range/ Requirements**	0 to 149°F
15)	**Weight**	27 lb
16)	**External Dimensions**	NR
17)	**Calibration Requirements/ Maintenance**	Factory calibration; automatic zero reference for noise reduction
18)	**Manufacturer**	OMEGA
19)	**Cost**	Series FMG 94 Magmeter, $2,562 for flowhead and electronics
20)	**Comments**	Wafer-type magnetic flowmeter; requires minimum fluid conductivity of 5 μS/cm; flange-type installation 2" pipe

Kobold

1)	**Analyte**	Flow/totalizer
2)	**Method**	Differential-pressure, direct-reading flowmeter
3)	**Range/Sensitivity**	Ranges from 15 to 100, 20 to 150, and 30 to 200 gpm for 2" pipe
4)	**Accuracy**	±3% FS horizontal flow; ±5% FS vertical flow
5)	**Repeatability/Life Expectancy**	NR
6)	**Zero Drift @ Constant Temp.**	NR
7)	**Zero Drift @ Ambient Temp.**	NR
8)	**Zero Noise @ Constant Temp.**	NR
9)	**External Power Supply**	24 VDC ±10%
10)	**Power Consumption**	NR
11)	**Output Signal**	4 to 20 mA
12)	**Electronic Response Time**	NR
13)	**Operating Temperature Range**	−30 to 212°F process temperature limit
14)	**Storage Temperature Range/ Requirements**	−30 to 120°F ambient temp limits (electrical specs)
15)	**Weight**	6.1 lb
16)	**External Dimensions**	6.8" H x 4.55" W x 3.33" D
17)	**Calibration Requirements/ Maintenance**	Factory calibration; on-site zero and span adjustments
18)	**Manufacturer**	Kobold
19)	**Cost**	RCM Industries direct-reading flowmeter, 2" SS, $975; electronics, $780
20)	**Comments**	Male pipe thread (MPT) fittings; no moving parts to be fouled; 10-pipe diameter minimum laminar flow prior to the flowmeter; valving recommended downstream of meter; compact

GLI International F1A11B2

1)	**Analyte**	Flow monitor/totalizer
2)	**Method**	Impeller-driven sensor
3)	**Range/Sensitivity**	12 to 380 gpm for 2" pipe in a PVC tee; 13 to 410 gpm in a PVDF tee
4)	**Accuracy**	1% of FS; ±1% FS linearity
5)	**Repeatability/Life Expectancy**	Repeatability of ±0.5% FS
6)	**Zero Drift @ Constant Temp.**	NR
7)	**Zero Drift @ Ambient Temp.**	NR
8)	**Zero Noise @ Constant Temp.**	NR
9)	**External Power Supply**	98 to 137 VAC standard; 24 VDC optional for digital flow monitor/totalizer
10)	**Power Consumption**	NR
11)	**Output Signal**	4 to 20 mA analog output; 24-VDC scalable pulse
12)	**Electronic Response Time**	3 sec to 90% on increasing flow; 2 sec on decreasing flow
13)	**Operating Temperature Range**	−22 to 122°F for electronics
14)	**Storage Temperature Range/ Requirements**	NR
15)	**Weight**	Electronics, ~3 lb; sensor, <1 lb
16)	**External Dimensions**	Electronics, 5.67" H x 5.67" W x 4.3" D; sensor ~5" L x ~1.5" D
17)	**Calibration Requirements/ Maintenance**	Factory calibration; on-site zero and span adjustments
18)	**Manufacturer**	GLI International, Inc.
19)	**Cost**	F1A11B2 PVDF tee-mount sensor, $360; 675 flow monitor/totalizer, $645; mounting hardware for 2", $510
20)	**Comments**	Impeller-type sensor; moving parts can be fouled; lower-cost sensor than vortex type

OMEGA FD-7000

1)	**Analyte**	Flow monitor/totalizer
2)	**Method**	Ultrasonic Doppler
3)	**Range/Sensitivity**	8-digit LCD in gallons per minute
4)	**Accuracy**	±2% full scale
5)	**Repeatability/Life Expectancy**	Linearity and repeatability of ±2% FS
6)	**Zero Drift @ Constant Temp.**	NR
7)	**Zero Drift @ Ambient Temp.**	NR
8)	**Zero Noise @ Constant Temp.**	NR
9)	**External Power Supply**	120 VAC or self-contained battery with charger
10)	**Power Consumption**	NR
11)	**Output Signal**	4 to 20 mA into 600 ohms
12)	**Electronic Response Time**	NR
13)	**Operating Temperature Range**	0 to 50°C for electronics, −40 to 82°C for fluids
14)	**Storage Temperature Range/ Requirements**	NR
15)	**Weight**	23 lb
16)	**External Dimensions**	17" L x 11" W x 8" H
17)	**Calibration Requirements/ Maintenance**	Factory calibration
18)	**Manufacturer**	OMEGA
19)	**Cost**	Model FD-7000, $6,500
20)	**Comments**	Noninvasive design; suitable for clear, low-viscosity liquids; for 1- to 60-inch pipes; includes a totalizer feature

Ellison Sensors Incorporated Model PR9500

1)	**Analyte**	Pressure
2)	**Method**	Silicon-on-sapphire sensor
3)	**Range/Sensitivity**	0 to 500 mbar or 0 to 1500 bar
4)	**Accuracy**	Resolution, 12-bit. Linearity and hysteresis, 0.3% full scale
5)	**Repeatability/Life Expectancy**	NR
6)	**Zero Drift @ Constant Temp.**	NR
7)	**Zero Drift @ Ambient Temp.**	NR
8)	**Zero Noise @ Constant Temp.**	NR
9)	**External Power Supply**	Transmitter and receiver, 10.5 to 30 VDC
10)	**Power Consumption**	NR
11)	**Output Signal**	4 to 20 mA analog output
12)	**Electronic Response Time**	Continuous sensing
13)	**Operating Temperature Range**	NR
14)	**Storage Temp. Range/ Requirements**	NR
15)	**Weight**	NR
16)	**External Dimensions**	Transmitter, 156 mm W x 60 mm H Receiver, 75 mm D x 120 mm W x 200 mm H
17)	**Calibration Requirements/ Maintenance**	Factory calibration. Zero and span adjustments.
18)	**Manufacturer**	Ellison Sensors, Incorporated (Protan)
19)	**Cost**	Model PR9500: transmitter, $975; receiver, $750
20)	**Comments**	Nonwired application, built-in low-power UHF radio transmitter with transmission range of 500 m, no license required

MODUS Series W

1)	**Analyte**	Pressure
2)	**Method**	Piezoresistive differential-pressure measurements.
3)	**Range/Sensitivity**	Differential pressure range, 0 to 200 psid, operating static pressure, 0 to 300 psi
4)	**Accuracy**	±0.5% of differential pressure range
5)	**Repeatability/Life Expectancy**	NR
6)	**Zero Drift @ Constant Temp.**	NR
7)	**Zero Drift @ Ambient Temp.**	NR
8)	**Zero Noise @ Constant Temp.**	NR
9)	**External Power Supply**	11 to 32 VDC
10)	**Power Consumption**	10 mA
11)	**Output Signal**	0 to 5 V, linear
12)	**Electronic Response Time**	Continuous
13)	**Operating Temperature Range**	32 to 122°F
14)	**Storage Temp. Range/ Requirements**	-20 to 160°F
15)	**Weight**	1.2 lb
16)	**External Dimensions**	3.56″ H x 4.54″ W x 2.18″ D
17)	**Calibration Requirements/ Maintenance**	Factory calibration traceable to National Institute of Standards and Technology (NIST). Field adjustments to zero and span according to supplied calibration data.
18)	**Manufacturer**	MODUS
19)	**Cost**	Series W: Base price, $395
20)	**Comments**	No moving parts, industrial applications, suitable for use in measuring the differential pressure across filters and pumps, takes the place of having two pressure transducers for filtration monitoring applications

NOSHOK, Inc. Series 100

1) Analyte	Pressure
2) Method	Thin-film sensor technology
3) Range/Sensitivity	0 to 100 psig
4) Accuracy	±0.5% full scale (standard); ±0.2% full scale (optional)
5) Repeatability/Life Expectancy	Repeatability, ±0.05% full scale; stability, ±0.2% full scale per year
6) Zero Drift @ Constant Temp.	NR
7) Zero Drift @ Ambient Temp.	Zero effect of ±0.03%/50°F
8) Zero Noise @ Constant Temp.	NR
9) External Power Supply	Input excitation of 12 to 30 VDC unregulated
10) Power Consumption	NR
11) Output Signal	4 to 20 mA, two-wire
12) Electronic Response Time	Less than 1 millisecond (between 10 and 90% full scale)
13) Operating Temperature Range	Compensated 32 to 175°F
14) Storage Temp. Range/ Requirements	-40 to 212°F.
15) Weight	NR
16) External Dimensions	4.5″ H x 1.07″ diameter
17) Calibration Requirements/ Maintenance	Factory calibration. Field adjustments to zero and span according to supplied calibration data.
18) Manufacturer	NOSHOK, Inc.
19) Cost	Series 100 pressure transducers: base price $226
20) Comments	Industrial-grade pressure transducers, reportedly very stable response, operating life of 100 million cycles

OMEGA Model PX120

1) Analyte	Pressure
2) Method	Thin-film sensor technology
3) Range/Sensitivity	0 to 500 psig
4) Accuracy	±1% FS (best-fit straight line [BFSL] includes linearity, hysteresis repeatability)
5) Repeatability/Life Expectancy	NR
6) Zero Drift @ Constant Temp.	NR
7) Zero Drift @ Ambient Temp.	Thermal zero effect of <±1% of full scale per 100°F
8) Zero Noise @ Constant Temp.	NR
9) External Power Supply	Excitation: 5 (±0.25) VDC @ 10 mA
10) Power Consumption	NR
11) Output Signal	50 ±1 mVDC with a sensitivity of 10 mV/V
12) Electronic Response Time	Less than 1 millisecond (between 10 to 90% full scale)
13) Operating Temperature Range	Compensated 32 to 175°F (0 to 80°C) Operating range: -40 to 257°F
14) Storage Temp. Range/ Requirements	-40 to 212°F.
15) Weight	3 oz
16) External Dimensions	1.88″ L x 1.0″ diameter
17) Calibration Requirements/ Maintenance	Factory calibration. Field adjustments to zero and span according to supplied calibration data.
18) Manufacturer	OMEGA
19) Cost	PX120 Series: $158
20) Comments	Industrial-grade pressure transducers, corrosion-resistant, operating life of 100 million cycles

OMEGA Model PX300

1) Analyte	Pressure
2) Method	Stainless steel diaphragm, silicone oil-filled semiconductor
3) Range/Sensitivity	0 to 500 psig
4) Accuracy	±0.25% full scale (includes linearity, hysteresis, repeatability)
5) Repeatability/Life Expectancy	NR
6) Zero Drift @ Constant Temp.	NR
7) Zero Drift @ Ambient Temp.	Thermal zero effect of 3% full scale maximum
8) Zero Noise @ Constant Temp.	NR
9) External Power Supply	Excitation: 5 (±0.25) VDC @ 10 mA
10) Power Consumption	NR
11) Output Signal	30 mV ±1 mV
12) Electronic Response Time	Less than 1 millisecond (between 10 to 90% full scale)
13) Operating Temperature Range	Compensated 30 to 160°F (−1 to 71°C) Operating range: 0 to 160°F
14) Storage Temp. Range/ Requirements	NR
15) Weight	4.6 oz
16) External Dimensions	3.375″ L x 1.0″ diameter
17) Calibration Requirements/ Maintenance	Factory calibration. Field adjustments to zero and span according to supplied calibration data
18) Manufacturer	OMEGA
19) Cost	PX300 Series: $180
20) Comments	Industrial-grade, general-purpose pressure sensor

OMEGA Models PX602 through PX612

1)	**Analyte**	Pressure
2)	**Method**	Thin-film polysilicon pressure sensor
3)	**Range/Sensitivity**	0 to 500 psig
4)	**Accuracy**	±0.4% full scale (includes linearity, hysteresis, repeatability)
5)	**Repeatability/Life Expectancy**	±0.05%
6)	**Zero Drift @ Constant Temp.**	NR
7)	**Zero Drift @ Ambient Temp.**	±0.04% full scale per °F
8)	**Zero Noise @ Constant Temp.**	NR
9)	**External Power Supply**	Excitation: 10 VDC
10)	**Power Consumption**	NR
11)	**Output Signal**	0 to 100 mV @ 10 VDC
12)	**Electronic Response Time**	1 millisecond
13)	**Operating Temperature Range**	Compensated −20 to 180°F. Operating range −55 to 195°F
14)	**Storage Temp. Range/ Requirements**	NR
15)	**Weight**	2.5 oz without the cable
16)	**External Dimensions**	3.78″ L x 1.058″ square cap
17)	**Calibration Requirements/ Maintenance**	Factory calibration. Field adjustments to zero and span according to supplied calibration data.
18)	**Manufacturer**	OMEGA
19)	**Cost**	PX602-PX612 Series: $198 + $10 snubber
20)	**Comments**	Industrial-grade, general-purpose pressure sensor, excellent long-term stability, psig ranges from 0 to 15 through 0 to 20,000 psig also available.

OMEGA Model PX425

1) Analyte	Pressure
2) Method	Molecular-bonded silicon strain gauge to sensing steel beam
3) Range/Sensitivity	Range from 0 to 15 psig through 0 to 6,000 psig
4) Accuracy	±0.19% span, includes linearity and hysteresis
5) Repeatability/Life Expectancy	Repeatability: maximum 0.05% span
6) Zero Drift @ Constant Temp.	NR
7) Zero Drift @ Ambient Temp.	±0.015% span/°C
8) Zero Noise @ Constant Temp.	NR
9) External Power Supply	Excitation: 10 VDC
10) Power Consumption	NR
11) Output Signal	30 mV ±0.3 mV
12) Electronic Response Time	1 millisecond
13) Operating Temperature Range	Compensated −65 to 248°F. Operating range −65 to 248°F
14) Storage Temp. Range/ Requirements	NR
15) Weight	6 oz
16) External Dimensions	2.5″ L x 1.17″ diameter
17) Calibration Requirements/ Maintenance	Factory calibration. Field adjustments to zero and span according to supplied calibration data. Rated for very high stability. Shock- and vibration-resistant.
18) Manufacturer	OMEGA
19) Cost	PX425: $495 + $10 snubber
20) Comments	Industrial-grade, general-purpose pressure sensor; excellent long-term stability; vibration-resistant.

OMEGA Model PX726

1)	**Analyte**	Pressure
2)	**Method**	Stainless steel diaphragm, silicone oil-filled semiconductor sensor
3)	**Range/Sensitivity**	Range from 0 to 10 psig through 0 to 5,000 psig
4)	**Accuracy**	±0.15% of calibrated span (includes linearity, hysteresis, repeatability)
5)	**Repeatability/Life Expectancy**	Repeatability: 0.05% span
6)	**Zero Drift @ Constant Temp.**	NR
7)	**Zero Drift @ Ambient Temp.**	NR
8)	**Zero Noise @ Constant Temp.**	NR
9)	**External Power Supply**	Excitation: 15 to 36 VDC
10)	**Power Consumption**	NR
11)	**Output Signal**	4 to 20 mA (two-wire)
12)	**Electronic Response Time**	50 milliseconds or 150 milliseconds with damping
13)	**Operating Temperature Range**	Process −40 to 220°F. Electrical −25 to 185°F
14)	**Storage Temp. Range/ Requirements**	NR
15)	**Weight**	1 lb 9 oz
16)	**External Dimensions**	5.82″ H x 4.4″ D
17)	**Calibration Requirements/ Maintenance**	Factory calibration. Field adjustments to zero and span according to supplied calibration data. 6:1 turndown on span. Complete zero suppression or elevation.
18)	**Manufacturer**	OMEGA
19)	**Cost**	PX726: $710 + $10 snubber
20)	**Comments**	Industrial flush-mount transmitter, NEMA 4x enclosure, ½″ NPTF conduit, screw terminals.

Setra Model 228-1/C228-1

1)	**Analyte**	Pressure
2)	**Method**	Capacitance differential-pressure sensor
3)	**Range/Sensitivity**	Differential-pressure range, 0 to 100 psid, maximum working pressure of 1,000 psig
4)	**Accuracy**	±0.25% full scale (linearity, hysteresis, repeatability)
5)	**Repeatability/Life Expectancy**	NR
6)	**Zero Drift @ Constant Temp.**	NR
7)	**Zero Drift @ Ambient Temp.**	Zero shift of <0.02% full scale per °F
8)	**Zero Noise @ Constant Temp.**	NR
9)	**External Power Supply**	24 VDC, 10 mA
10)	**Power Consumption**	NR
11)	**Output Signal**	0 to 5 VDC
12)	**Electronic Response Time**	30 to 50 milliseconds
13)	**Operating Temperature Range**	0 to 175°F
14)	**Storage Temp. Range/ Requirements**	−65 to 250°F
15)	**Weight**	~2 lb
16)	**External Dimensions**	3.05″ D x 2.7″ diameter
17)	**Calibration Requirements/ Maintenance**	Factory-zeroed after application of 300 psi to high-pressure port. Needs to be rezeroed after application of high pressure.
18)	**Manufacturer**	Setra
19)	**Cost**	Model 228-1/C228-1: $870
20)	**Comments**	Fast response, gas or liquid compatible, low line pressure effects, applicable for determining pressure drop across filters under high working pressures.

Setra Model C230

1) Analyte	Pressure
2) Method	Capacitance differential-pressure sensor
3) Range/Sensitivity	Differential pressure range, 0 to 100 psid, maximum working pressure of 250 psig
4) Accuracy	±0.25% full scale at constant temperature (linearity, hysteresis repeatability)
5) Repeatability/Life Expectancy	Repeatability: ±0.05% full scale
6) Zero Drift @ Constant Temp.	NR
7) Zero Drift @ Ambient Temp.	Zero shift of <±0.02% full scale per °F
8) Zero Noise @ Constant Temp.	NR
9) External Power Supply	11 to 30 VDC
10) Power Consumption	NR
11) Output Signal	4 to 20 mA (20-wire)
12) Electronic Response Time	30 to 50 milliseconds
13) Operating Temperature Range	0 to 175°F
14) Storage Temp. Range/ Requirements	-65 to 250°F
15) Weight	14.4 oz
16) External Dimensions	3″ H x 2.84″ D x 2.15″ W
17) Calibration Requirements/ Maintenance	Factory prezeroed after maximum overload pressure to high-pressure port. Zero may shift with high differential overpressure.
18) Manufacturer	Setra
19) Cost	Model C230: $250
20) Comments	Fast response, gas or liquid compatible, low line pressure effects, applicable for determining pressure drop across filters under moderate (<250 psig) working pressures.

Kari Model M2HL

1)	**Analyte**	Level measurement
2)	**Method**	Cable-suspended float switches
3)	**Range/Sensitivity**	Standard cable length of 16'
4)	**Accuracy**	10" minimum switching differential, 40" maximum
5)	**Repeatability/Life Expectancy**	NR
6)	**Zero Drift @ Constant Temp.**	NR
7)	**Zero Drift @ Ambient Temp.**	NR
8)	**Zero Noise @ Constant Temp.**	NR
9)	**External Power Supply**	Maximum contact rating: 250 VDC
10)	**Power Consumption**	NR
11)	**Output Signal**	Contact closure switch
12)	**Electronic Response Time**	NR
13)	**Operating Temperature Range**	Maximum liquid temperature: 140°F
14)	**Storage Temp. Range/ Requirements**	NR
15)	**Weight**	Standard type float: 0.5 lb
16)	**External Dimensions**	Float: 3" diameter, 4.85" L
17)	**Calibration Requirements/ Maintenance**	Contact switch is factory-adjusted; high-alarm, low-alarm settings adjusted in the field by varying the length of cable between the weight and the float
18)	**Manufacturer**	Kari supplied by Scientific Technology Incorporated
19)	**Cost**	Model: M2HL, $127 for the float and cable
20)	**Comments**	Need minimum specific gravity of 0.95. Simple indicator of a "high" or "low" tank level. Contact easily integrated into an electronics package for remote alarms. Additional units available that can deliver contacts for four levels.

Kobold Model NM-298

1) Analyte	Level measurement
2) Method	Mechanical with reed contacts
3) Range/Sensitivity	Range of 0" to 36", with additional 4" increments available
4) Accuracy	±0.6"
5) Repeatability/Life Expectancy	NR
6) Zero Drift @ Constant Temp.	NR
7) Zero Drift @ Ambient Temp.	NR
8) Zero Noise @ Constant Temp.	NR
9) External Power Supply	Sensor voltage, 20 VDC max. Transmitter, 24 VDC
10) Power Consumption	NR
11) Output Signal	Options from the meter, 0 to 20, 4 to 20 mA or 0 to 5, 0 to 10 VDC
12) Electronic Response Time	NR
13) Operating Temperature Range	-4° to 250°F for sensor, 0°to 140°F for the controller/transmitter
14) Storage Temp. Range/ Requirements	NR
15) Weight	NR
16) External Dimensions	Float: 1.7" diameter, 1.8" L Shaft: 0.5" diameter, up to 6.6' L
17) Calibration Requirements/ Maintenance	Little or no maintenance needed
18) Manufacturer	Kobold
19) Cost	Model: NM-298 transducer, $619; 36" shaft, $304; model type DFM signal conditioner, $222
20) Comments	Liquid has to have a minimum specific gravity of 0.8. In conjunction with the controller transmitter it may be possible to use this to refill tank.

OMEGA Model LVR-30

1)	**Analyte**	Level measurement
2)	**Method**	Voltage divider network, magnetic float/reed switches
3)	**Range/Sensitivity**	Measurements from 0' to 6'
4)	**Accuracy**	0.4% of full scale or ±0.5"
5)	**Repeatability/Life Expectancy**	NR
6)	**Zero Drift @ Constant Temp.**	NR
7)	**Zero Drift @ Ambient Temp.**	NR
8)	**Zero Noise @ Constant Temp.**	NR
9)	**External Power Supply**	10 to 40 VDC
10)	**Power Consumption**	NR
11)	**Output Signal**	4 to 20 mA output
12)	**Electronic Response Time**	Continuous readout
13)	**Operating Temperature Range**	5°F to 160°F
14)	**Storage Temp. Range/ Requirements**	Storage temperature of −40°F to 212°F
15)	**Weight**	5 lb + 0.25 lb per foot of measurement length
16)	**External Dimensions**	2" NPTM fitting
17)	**Calibration Requirements/ Maintenance**	Factory-calibrated
18)	**Manufacturer**	OMEGA
19)	**Cost**	Model: LVR-30, $458; Model DPF50 process meter, $160
20)	**Comments**	Chemically resistant materials of 316 stainless steel. Assumes liquid has a specific gravity of at least 0.75. Simple float on a rod design. Immune to density or dielectric changes. Low-cost unit.

GLI International Model 4000L

1) Analyte	Level measurement
2) Method	Conductance on the electrode
3) Range/Sensitivity	Measurements from 0' to 19.5', in 1-inch increments
4) Accuracy	Linearity of 0.5% full scale
5) Repeatability/Life Expectancy	NR
6) Zero Drift @ Constant Temp.	NR
7) Zero Drift @ Ambient Temp.	NR
8) Zero Noise @ Constant Temp.	NR
9) External Power Supply	18 to 35 VDC or 115 VAC
10) Power Consumption	NR
11) Output Signal	4 to 20 mA transmitter, integral or remote mount
12) Electronic Response Time	0 to 10-sec response output damping
13) Operating Temperature Range	Maximum temperature 450°F for Teflon™, 250°F for Kynar™, 160°F for polypropylene
14) Storage Temp. Range/ Requirements	NR
15) Weight	In aluminum enclosure, 3.6 lb, in PVC enclosure, 3.4 lb
16) External Dimensions	Electrode diameter, 3/8"; lengths 0' to 12', 12' to 19.5'
17) Calibration Requirements/ Maintenance	Factory-calibrated
18) Manufacturer	GLI International, Inc.
19) Cost	Model: 4000L probe, ~$200; Model 690L transmitter, $590
20) Comments	Chemically resistant materials of 316 stainless steel or Teflon™. Not recommended for use in materials which tend to form a conductive film on surface of sensing element. Needs Model 690L module to get signal from the sensor.

GLI International Model 4300L

1)	**Analyte**	Level measurement
2)	**Method**	Conductance on the electrode, flexible cable
3)	**Range/Sensitivity**	Measurements from 0' to 12', in 1-inch increments
4)	**Accuracy**	Linearity of 0.5% full scale
5)	**Repeatability/Life Expectancy**	NR
6)	**Zero Drift @ Constant Temp.**	NR
7)	**Zero Drift @ Ambient Temp.**	NR
8)	**Zero Noise @ Constant Temp.**	NR
9)	**External Power Supply**	18 to 35 VDC or 115 VAC
10)	**Power Consumption**	NR
11)	**Output Signal**	4 to 20 mA transmitter, integral or remote mount
12)	**Electronic Response Time**	0 to 10-sec response output damping
13)	**Operating Temperature Range**	Maximum temperature, 350°F for Teflon™-coated cable
14)	**Storage Temp. Range/ Requirements**	NR
15)	**Weight**	In aluminum enclosure, 3.6 lb; in PVC enclosure, 3.4 lb
16)	**External Dimensions**	Electrode diameter, ~3/8"; length 0' to 12'
17)	**Calibration Requirements/ Maintenance**	Factory-calibrated
18)	**Manufacturer**	GLI International, Inc.
19)	**Cost**	Model: 4300L, ~$300; Model: 690L transmitter, $590
20)	**Comments**	Chemically resistant materials of 316 stainless steel or Teflon™. Not recommended for use in materials which tend to form a conductive film on surface of sensing element. Need Model 690L module to get signal from the sensor.

In-Situ, Inc. Model PXD-260

1)	**Analyte**	Level measurement
2)	**Method**	Integrated silicon strain gauge bridge, pressure transducer
3)	**Range/Sensitivity**	Range from 0' to 578' depending on unit selected
4)	**Accuracy**	±0.15% of range-linear coefficients
5)	**Repeatability/Life Expectancy**	NR
6)	**Zero Drift @ Constant Temp.**	NR
7)	**Zero Drift @ Ambient Temp.**	NR
8)	**Zero Noise @ Constant Temp.**	NR
9)	**External Power Supply**	9 to 15 VDC
10)	**Power Consumption**	NR
11)	**Output Signal**	4 to 20 mA overpressure range
12)	**Electronic Response Time**	Instantaneous
13)	**Operating Temperature Range**	7° to 176°F
14)	**Storage Temp. Range/ Requirements**	-40° to 257°F
15)	**Weight**	1.5 lb
16)	**External Dimensions**	Probe head is 1" diameter by 10.9" L
17)	**Calibration Requirements/ Maintenance**	Factory calibration provided, low maintenance because typically used for in situ measurements of fluctuations in groundwater
18)	**Manufacturer**	In-Situ, Inc.
19)	**Cost**	PXD-260 ~$1,100
20)	**Comments**	Made with 316 stainless steel and Viton components. Very accurate results when used to monitor groundwater fluctuations.

Kobold Model NEO-5000

1)	**Analyte**	Level measurement
2)	**Method**	Ultrasonic
3)	**Range/Sensitivity**	Measure depth range of 0.5' to 24.5' from sensor face
4)	**Accuracy**	±0.25% of full scale
5)	**Repeatability/Life Expectancy**	Repeatability of ±0.125"
6)	**Zero Drift @ Constant Temp.**	NR
7)	**Zero Drift @ Ambient Temp.**	NR
8)	**Zero Noise @ Constant Temp.**	NR
9)	**External Power Supply**	24 VDC, external power
10)	**Power Consumption**	200 mA maximum
11)	**Output Signal**	4 to 20 mA
12)	**Electronic Response Time**	NR
13)	**Operating Temperature Range**	-40° to 148°F
14)	**Storage Temp. Range/ Requirements**	NR
15)	**Weight**	NR
16)	**External Dimensions**	3" L x 2" diameter
17)	**Calibration Requirements/ Maintenance**	Onboard LED display for on-site calibration, automatic temperature compensation. Artificial intelligence to distinguish real echoes.
18)	**Manufacturer**	Kobold
19)	**Cost**	Model: NEO-5000, $850
20)	**Comments**	Built-in relay may be used to control tank fill/empty operations or as an alarm or fault detection.

OMEGA Model LVU-300

1) Analyte	Level measurement
2) Method	Continuous, ultrasonic sensor
3) Range/Sensitivity	Measurements, 24.5' max from sensor face
4) Accuracy	±0.25% of full scale
5) Repeatability/Life Expectancy	Repeatability of ±0.125"
6) Zero Drift @ Constant Temp.	NR
7) Zero Drift @ Ambient Temp.	NR
8) Zero Noise @ Constant Temp.	NR
9) External Power Supply	14 to 36 VDC
10) Power Consumption	NR
11) Output Signal	4 to 20 mA + 12 amp relay @ 110 to 220 VAC
12) Electronic Response Time	Continuous sensing
13) Operating Temperature Range	-5°F to 140°F
14) Storage Temp. Range/ Requirements	NR
15) Weight	NR
16) External Dimensions	Sensor housing 2 x 2.2", electrical housing, 2.8" x 4.5"
17) Calibration Requirements/ Maintenance	Factory-calibrated
18) Manufacturer	OMEGA
19) Cost	Model: LVU-300, $895; Signal controller: LVCN-51, $425
20) Comments	Noncontact measurements; automatic temperature compensation; single relay for high/low, auto fill/auto empty; remote mount proportional controller compatible (Model LVCN-51) permits alarms, multiple setpoints.

Kobold Model NEA-2

1) Analyte	Level measurement
2) Method	Acoustic
3) Range/Sensitivity	Measurements from 0.5' to 4'
4) Accuracy	In air, ±1% span
5) Repeatability/Life Expectancy	Repeatability of ±0.125"
6) Zero Drift @ Constant Temp.	NR
7) Zero Drift @ Ambient Temp.	NR
8) Zero Noise @ Constant Temp.	NR
9) External Power Supply	12 to 36 VDC
10) Power Consumption	NR
11) Output Signal	4 to 20 mA
12) Electronic Response Time	Ping rate of 5 per second
13) Operating Temperature Range	-4° to 140°F
14) Storage Temp. Range/ Requirements	NR
15) Weight	NR
16) External Dimensions	Sensor head, 2.8" diameter x 4.9" L. Waveguide tube, 48"
17) Calibration Requirements/ Maintenance	Fixed offset. Automatic temperature compensation. Tuned to waveguide tube length.
18) Manufacturer	Kobold
19) Cost	Model: NEA-2, $449
20) Comments	Not affected by vapors, turbulence in a tank, low interference problems because the wave emitted travels in the protective waveguide as opposed to in open-air ultrasonic sensors.

INOTEK Technologies, Inc. Model 8206

1)	**Analyte**	Level measurement
2)	**Method**	Nonintrusive capacitance switch
3)	**Range/Sensitivity**	Measures upper or lower limit in a tank
4)	**Accuracy**	±1 mm in water
5)	**Repeatability/Life Expectancy**	NR
6)	**Zero Drift @ Constant Temp.**	NR
7)	**Zero Drift @ Ambient Temp.**	NR
8)	**Zero Noise @ Constant Temp.**	NR
9)	**External Power Supply**	12 to 36 VDC
10)	**Power Consumption**	Dry, 5 mA; wet, 19 mA
11)	**Output Signal**	4 to 20 mA output
12)	**Electronic Response Time**	Continuous monitoring
13)	**Operating Temperature Range**	-40° to 158°F
14)	**Storage Temp. Range/ Requirements**	NR
15)	**Weight**	NR
16)	**External Dimensions**	2.5" x 1.7" x 1.5"
17)	**Calibration Requirements/ Maintenance**	None required because the sensor acts as an on/off type switch
18)	**Manufacturer**	INOTEK Technologies, Inc.
19)	**Cost**	Model 8206, $195
20)	**Comments**	Sensor can be located on the outside of a nonmetallic tank that has up to a 1" thickness. This type of sensor could be used to indicate that a tank is full or empty.

OMEGA Model LVC-101

1)	**Analyte**	Level measurement
2)	**Method**	Intrusive radiofrequency capacitance sensor
3)	**Range/Sensitivity**	Based on cable length (standard 8') and sensor's position in tank
4)	**Accuracy**	±1 mm in water
5)	**Repeatability/Life Expectancy**	±0.5 mm
6)	**Zero Drift @ Constant Temp.**	NR
7)	**Zero Drift @ Ambient Temp.**	NR
8)	**Zero Noise @ Constant Temp.**	NR
9)	**External Power Supply**	12 to 36 VDC input
10)	**Power Consumption**	NR
11)	**Output Signal**	When sensor is dry, 4 mA output; when wet, 20 mA output
12)	**Electronic Response Time**	NR
13)	**Operating Temperature Range**	Maximum liquid temperature: 194°F
14)	**Storage Temp. Range/ Requirements**	NR
15)	**Weight**	NR
16)	**External Dimensions**	3" x 3/4"
17)	**Calibration Requirements/ Maintenance**	Factory calibration ensures reliability with a wide range of polar liquids
18)	**Manufacturer**	OMEGA
19)	**Cost**	Model: LVC-101, $185. Model: LVM-13, 6' track mount, $120
20)	**Comments**	The sensor is impervious to changes in temperature, pressure, light, particulates, liquid color, or liquid density.

OMEGA Model LV-5200

1)	**Analyte**	Level measurement
2)	**Method**	Radiofrequency capacitive level measurements
3)	**Range/Sensitivity**	Maximum length of 12'
4)	**Accuracy**	When using the LV-5900 transmitter, ±0.5% full scale
5)	**Repeatability/Life Expectancy**	Repeatability of ±0.5% of full scale
6)	**Zero Drift @ Constant Temp.**	NR
7)	**Zero Drift @ Ambient Temp.**	NR
8)	**Zero Noise @ Constant Temp.**	NR
9)	**External Power Supply**	18 to 35 VDC (std)
10)	**Power Consumption**	1 W @ 24 VDC
11)	**Output Signal**	4 to 20 mA output, field-selectable direct or reverse acting
12)	**Electronic Response Time**	Continuous sensing
13)	**Operating Temperature Range**	For Teflon™-coated sensor, 450°F
14)	**Storage Temp. Range/ Requirements**	NR
15)	**Weight**	Sensor, 1 lb for gland + 0.6 lb/ft for Teflon™. Transmitter, 3.6 lb
16)	**External Dimensions**	Sensor gland, 3/4" NPT; transmitter, 6.9 x 4.3"
17)	**Calibration Requirements/ Maintenance**	Transmitter requires span range and zero range adjustments when deployed
18)	**Manufacturer**	OMEGA. Model LV-5200 sensing element + LV-5900 transmitter
19)	**Cost**	Sensor, \$180. Transmitter, \$655 plus ~\$600 for accessory parts
20)	**Comments**	Sensor element has a conductive film-resistant coating. The transmitter can be located to or externally from the sensor, is weatherproof, provides compatible signal for RTS use.

VWR MANOSTAT

1)	**Parameter**	Pump
2)	**Principle of Operation**	Peristaltic, two-roller pump
3)	**Range/Sensitivity**	3 to 360 mL per minute
4)	**Accuracy**	NR
5)	**Repeatability/Life Expectancy**	NR
6)	**Zero Drift @ Constant Temp.**	NR
7)	**Zero Drift @ Ambient Temp.**	NR
8)	**Zero Noise @ Constant Temp.**	NR
9)	**External Power Supply**	115 V, 60 Hz, 1 amp
10)	**Power Consumption**	NR
11)	**Output Signal**	NA
12)	**Electronic Response Time**	NR
13)	**Operating Temperature Range**	NR
14)	**Storage Temp. Range/ Requirements**	NR
15)	**Weight**	5 lb
16)	**External Dimensions**	5" L x $6^9/_{16}$" D x $5^5/_{16}$" H
17)	**Calibration Requirements/ Maintenance**	None
18)	**Manufacturer**	MANOSTAT, distributed by VWR
19)	**Cost**	$198
20)	**Comments**	Designed to provide constant flow of liquid with a wide range of viscosities or gases. Can also be used for metering applications. 25 psi continuous and 40 psi intermittent delivery pressure.

KNF Neuberger, Inc. NF30KN

1)	**Parameter**	Pump
2)	**Principle of Operation**	Microdiaphragm pump
3)	**Range/Sensitivity**	Maximum flow 250 mL per minute
4)	**Accuracy**	NR
5)	**Repeatability/Life Expectancy**	NR
6)	**Zero Drift @ Constant Temp.**	NR
7)	**Zero Drift @ Ambient Temp.**	NR
8)	**Zero Noise @ Constant Temp.**	NR
9)	**External Power Supply**	115 V, 60 Hz
10)	**Power Consumption**	0.11 amp
11)	**Output Signal**	NA
12)	**Electronic Response Time**	NA
13)	**Operating Temperature Range**	Maximum ambient temperature of 105°F, water temperature 14 to 176°F
14)	**Storage Temp. Range/ Requirements**	NR
15)	**Weight**	1.3 lb
16)	**External Dimensions**	2.9" L x 1.2" D x 2.25" H
17)	**Calibration Requirements/ Maintenance**	None
18)	**Manufacturer**	KNF Neuberger, Inc.
19)	**Cost**	Model NF30KN, $149
20)	**Comments**	Single-head micro pump, maximum continuous pressure of 15 psig, maximum suction head, 20' H_2O

FMI Model QV-RH1-CKC-W

1)	**Parameter**	Pump
2)	**Principle of Operation**	Valveless piston pump
3)	**Range/Sensitivity**	0 to 180 mL per minute
4)	**Accuracy**	±1%
5)	**Repeatability/Life Expectancy**	NR
6)	**Zero Drift @ Constant Temp.**	NR
7)	**Zero Drift @ Ambient Temp.**	NR
8)	**Zero Noise @ Constant Temp.**	NR
9)	**External Power Supply**	115 V
10)	**Power Consumption**	NR
11)	**Output Signal**	4 to 20 mA
12)	**Electronic Response Time**	NR
13)	**Operating Temperature Range**	39 to 104°F
14)	**Storage Temp. Range/ Requirements**	NR
15)	**Weight**	8 lb
16)	**External Dimensions**	$7^1/_8$" L x 3" D x 3" H
17)	**Calibration Requirements/ Maintenance**	None
18)	**Manufacturer**	FMI (Fluid Metering, Inc.)
19)	**Cost**	Model QV-RH1-CKC-W, $1,095
20)	**Comments**	Positive-displacement, high-pressure (100 psig) pump. Piston may be susceptible to scoring from precipitated salts.

OMEGA Series SV-700

1)	**Parameter**	Valve
2)	**Principle of Operation**	Pneumatic diaphragm valve
3)	**Range/Sensitivity**	22 and 232 gpm for 0.5-inch and 2-inch pipes, respectively
4)	**Accuracy**	NR
5)	**Repeatability/Life Expectancy**	NR
6)	**Zero Drift @ Constant Temp.**	NR
7)	**Zero Drift @ Ambient Temp.**	NR
8)	**Zero Noise @ Constant Temp.**	NR
9)	**External Power Supply**	NA
10)	**Power Consumption**	NA
11)	**Output Signal**	NA
12)	**Electronic Response Time**	NA
13)	**Operating Temperature Range**	14 to 140°F
14)	**Storage Temp. Range/ Requirements**	NR
15)	**Weight**	0.5-inch orifice, 1.32 lb; 2-inch orifice, 14.57 lb
16)	**External Dimensions**	NR
17)	**Calibration Requirements/ Maintenance**	None
18)	**Manufacturer**	OMEGA
19)	**Cost**	Model SV-701 with 0.5-inch orifice, $222; SV-706 with 2-inch orifice, $490
20)	**Comments**	Additional components needed to operate this valve are an air compressor with a minimum of 80 psi pressure delivery, but less than 145 psi. Grainger supplies units of this type for approximately $250. Also needed are an electrical position indicator (SV7-012, $220) and a stroke limiter (SV7-SL2, $73). Will also need a controllable solenoid to deliver pressure to the valve.

OMEGA Series SV-350

1)	**Parameter**	Valve
2)	**Principle of Operation**	Electrically activated solenoid valve
3)	**Range/Sensitivity**	Varies for opening size and pressure
4)	**Accuracy**	NR
5)	**Repeatability/Life Expectancy**	NR
6)	**Zero Drift @ Constant Temp.**	NR
7)	**Zero Drift @ Ambient Temp.**	NR
8)	**Zero Noise @ Constant Temp.**	NR
9)	**External Power Supply**	120 VAC
10)	**Power Consumption**	NR
11)	**Output Signal**	NA
12)	**Electronic Response Time**	Opening time, 100 to 1,000 msec; closing time, 70 to 4,000 msec
13)	**Operating Temperature Range**	Up to 130°F
14)	**Storage Temp. Range/ Requirements**	NR
15)	**Weight**	NR
16)	**External Dimensions**	NR
17)	**Calibration Requirements/ Maintenance**	None
18)	**Manufacturer**	OMEGA
19)	**Cost**	Model SV-351, 0.5-inch NPT, $98; SV-359, 2-inch NPT, $560
20)	**Comments**	Can be operated on 12 or 24 VDC. Will require contact closure electronics to deliver power to activate the solenoid.

D

COTS Data Acquisition Options

The DAQ products listed here represent several types of DAQ systems, including DAQ boards, dataloggers, and mounting racks with I/O modules. They appear in the same order as on Table 6-1 in the text.

This listing is provided solely for informational purposes. Inclusion in this appendix does not constitute recommendation of the companies listed or endorsement for use of their products. Reasonable efforts have been made to ensure that the information provided is up to date.

Acromag, AVME9510-1

1) Principle of Operation	DAQ board interfaces digital and analog I/O signals to VME bus
2) Interface with Telemetry Device	RS232
3) Number of Bits Resolution	12-bit A/D, 32-μS conversion rate
4) Analog I/O	±5 V, ±10 V, 0- to 10-V input and outputs
5) Digital I/O	32 points of bidirectional I/O, TTL/CMOS-compatible, 0- to 5-V and 0- to 30-V input
6) Number of Input Channels	16 differential or 32 single-ended high-level inputs; 2 output channels
7) Manufacturer/Model	Acromag Model AVME9510-1
8) Cost	NR
9) Comments	Has both analog and digital I/O; problem with one channel may require replacement of whole board

Campbell Scientific, Inc., CR23X

1) Principle of Operation	Self-contained, rugged datalogger with nonvolatile data storage, battery backup, and real-time clock
2) Interface with Telemetry Device	RS232 and CSI standard 9-pin
3) Number of Bits Resolution	15-bit
4) Analog I/O	Inputs ±5 VDC, ±1 VDC, ±200 mVDC, ±50 mVDC, ±10 mVDC
5) Digital I/O	8 I/O ports
6) Number of Input Channels	12 differential or 24 single-ended analog inputs; 8 digital I/O ports
7) Manufacturer/Model	Campbell Scientific, Inc., Model CR23X
8) Cost	$1,090
9) Comments	PC208W datalogger support software for Windows®; also provides remote telemetry options (RF, phone, cellular, satellite)

Data Electronics, Datalogger DT500

1) Principle of Operation	Microprocessor-based, or main-powered devices designed to operate in any environment, independent of host computer
2) Interface with Telemetry Device	RS232 full duplex
3) Number of Bits Resolution	15-bit/1 μV; accuracy 0.15% of full scale
4) Analog I/O	Input 4 to 20 mA, volts
5) Digital I/O	NR
6) Number of Input Channels	10 differential channels, 4 inputs per channel
7) Manufacturer/Model	Data Electronics, Datalogger DT500
8) Cost	$2,575 with battery and software
9) Comments	Signals do not have to be conditioned; 1 MB storage space (340,000 readings) in PCMCIA card

Daytronic, 10KU-EL

1) Principle of Operation	High-speed front-end data acquisition and logging
2) Interface with Telemetry Device	RS232
3) Number of Bits Resolution	16-bit
4) Analog I/O	Any sensor input
5) Digital I/O	None
6) Number of Input Channels	160 analog inputs
7) Manufacturer/Model	Daytronic, Model 10KU-EL
8) Cost	$2,195
9) Comments	Datalogger interfaces with any computer/operating system and programming language; Labview software

E. Clark & Associates, ALMEMO 2290-8 Measured Value Logger

1)	**Principle of Operation**	Automatic acquisition datalogger
2)	**Interface with Telemetry Device**	RS232
3)	**Number of Bits Resolution**	16-bit A/D
4)	**Analog I/O**	Input signals
5)	**Digital I/O**	Frequencies and pulses
6)	**Number of Input Channels**	4 analog inputs, 1 digital input
7)	**Manufacturer/Model**	E. Clark & Associates, Model ALMEMO 2290-8 Measured Value Logger
8)	**Cost**	NR
9)	**Comments**	Different sensor types can be connected without special programming

Handar, 555 Universal Data Acquisition System

1)	**Principle of Operation**	Datalogger with DAQ board and modem/radio options
2)	**Interface with Telemetry Device**	SDI-12 Smart Sensor interface
3)	**Number of Bits Resolution**	14-bit
4)	**Analog I/O**	Input −2.5 to 5 VDC, −25 to 50 mVDC
5)	**Digital I/O**	Programmable for control
6)	**Number of Input Channels**	16 analog inputs, 8 digital inputs, 8 digital outputs
7)	**Manufacturer/Model**	Handar, Model 555 Universal Data Acquisition System
8)	**Cost**	NR
9)	**Comments**	NEMA-4 enclosure houses 12-V battery, choice of telemetry or data storage options; experienced in remote telemetry; system has been used for air and water quality management

In-Situ, Inc., Well Sentinel 3100

1)	**Principle of Operation**	Four-channel datalogger; can read and store 500,000 readings
2)	**Interface with Telemetry Device**	RS232 serial port, upgradable to RS422/485; IBM 9-pin
3)	**Number of Bits Resolution**	0.01% full scale; accuracy ±0.1% full scale
4)	**Analog I/O**	4 to 20 mA input
5)	**Digital I/O**	NR
6)	**Number of Input Channels**	4 external, 2 internal
7)	**Manufacturer/Model**	In-Situ, Inc., Well Sentinel 3100
8)	**Cost**	$3,800
9)	**Comments**	Nonvolatile 1024K memory; fits in 4-inch Schedule 80 PVC well casing, adaptable to 2-inch wells; 967K data storage; used to monitor water quality/process parameters

ITT Barton, DataScan DS40 Measurement Telemetry

1)	**Principle of Operation**	Standalone datalogger with internal 2400-baud modem
2)	**Interface with Telemetry Device**	RS232 full duplex; RS485, half-duplex; serial board
3)	**Number of Bits Resolution**	16-bit; ±0.1% of full scale for 4 to 20 mA inputs
4)	**Analog I/O**	4 to 20 mA input and output
5)	**Digital I/O**	None
6)	**Number of Input Channels**	6 analog inputs, 2 analog outputs
7)	**Manufacturer/Model**	ITT Barton, DataScan DS40 Measurement Telemetry
8)	**Cost**	NR
9)	**Comments**	Used for DAQ and telemetry; NEMA-4/IP66 enclosure

Logic Beach, Inc., Minilogger

1) Principle of Operation	Self-contained portable datalogging system with system microprocessor, data storage memory, A/D converter, and signal-conditioning circuitry for inputs and outputs
2) Interface with Telemetry Device	RS232
3) Number of Bits Resolution	12-bit
4) Analog I/O	Several current and voltage inputs; 100 mA, 5-VDC output
5) Digital I/O	Current-limited TTL digital outputs
6) Number of Input Channels	24 analog, 1 digital
7) Manufacturer/Model	Logic Beach, Inc., Minilogger
8) Cost	$1,200 for Minilogger, $1,800 for Modulogger
9) Comments	Reprogramming of datalogger possible with remote PC; Hyperware software provided; has control functions; battery backs up static random access memory (SRAM) (15,000 samples)

OMEGA, OM-220

1) Principle of Operation	Remote-controlled portable datalogging system
2) Interface with Telemetry Device	RS232, RS485
3) Number of Bits Resolution	12-bit dual-slope integrating
4) Analog I/O	Input 4 to 20 mA; various sensor inputs; two programmable output voltages
5) Digital I/O	NR
6) Number of Input Channels	4 modules maximum, 4 channels per module; can select digital or analog modules
7) Manufacturer/Model	OMEGA, Model OM-220
8) Cost	$1,350
9) Comments	Inexpensive, versatile

Onset Computer Corporation, Tattletale 5F

1) Principle of Operation	Specialty computer boards with integrated hardware and software components; analog inputs from sensors, digital I/O, nonvolatile program and data storage options
2) Interface with Telemetry Device	RS232
3) Number of Bits Resolution	12-bit
4) Analog I/O	0- to 5-V input
5) Digital I/O	Programmable for control
6) Number of Input Channels	14 digital I/O, 8 analog I/O
7) Manufacturer/Model	Onset Computer Corporation, Tattletale 5F
8) Cost	$500
9) Comments	Small, light, inexpensive; program stored in Flash EEPROM, embedded intelligence; used for remote monitoring in aerospace, oceanography, and several other applications; 480 kB RAM

OPTO, 22 PB16AI

1) Principle of Operation	Analog I/O mounting rack accommodates 16 I/O modules and interfaces with brain board via a 50-pin header connector
2) Interface with Telemetry Device	RS485
3) Number of Bits Resolution	12-bit
4) Analog I/O	4 to 20 mA, 0 to 5 VDC
5) Digital I/O	Not applicable
6) Number of Input Channels	16
7) Manufacturer/Model	OPTO 22, Model PB16AI
8) Cost	$1,000
9) Comments	Suitable I/O modules have to be selected; modules easily replaced, multiple source

OPTO 22, PB16AH

1) Principle of Operation	Digital I/O mounting rack accommodates 16 I/O modules and interfaces with brain board via a 50-pin card edge connector
2) Interface with Telemetry Device	Six 32-screw terminals; 50-pin male card edge connector
3) Number of Bits Resolution	12-bit
4) Analog I/O	Not applicable
5) Digital I/O	Input and output available for control
6) Number of Input Channels	16
7) Manufacturer/Model	OPTO 22, Model PB16AH
8) Cost	$1,000
9) Comments	Suitable I/Os have to be selected; modules easily replaced, multiple source

E

COTS Data Transmission Systems

The data transmission/remote telemetry systems listed here represent several types of telephone, cellular, radiofrequency, and satellite telecommunications systems. They appear in the same order as on Table 7-2 in the text.

This listing is provided solely for informational purposes. Inclusion in this appendix does not constitute recommendation of the companies listed or endorsement for use of their products. Reasonable efforts have been made to ensure that the information provided is up to date.

Synetcom, Model RTU-1

1) **Principle of Operation**	Telephone modem, spread-spectrum RF, cellular, two-way radio
2) **Interface with DAQ System**	NR
3) **Major Components**	1200/2400 baud MSK modem for RF
4) **Transmission Range**	NR
5) **Transmission Rate**	NR
6) **Frequency Range**	Synthesized radios for 450 and 900 MHz Narrow-band radios from 150 to 980 MHz Spread spectrum radios for 900 MHz
7) **Manufacturer/Model**	Synetcom Digital, Inc., Model RTU-1 datalogger/remote controller with radio telemetry
8) **Cost**	$750 ($1,500 with logging and telemetry software)
9) **Comments**	Has cellular-compatible modem; has datalogging and remote telemetry capability.

ITT Barton, Model DS40

1) **Principle of Operation**	Analog data inputs; transmits data via internal modem at 2400 baud
2) **Interface with DAQ System**	Modem interface: RS232, full duplex; RS485 half duplex
3) **Major Components**	NR
4) **Transmission Range**	NR
5) **Transmission Rate**	NR
6) **Frequency Range**	NR
7) **Manufacturer/Model**	ITT Barton, Model DS40
8) **Cost**	NR
9) **Comments**	Used for DAQ and remote telemetry

Dancer Communications, Inc., Messenger 500 Series

1)	**Principle of Operation**	Built-in datalogger, nonvolatile memory, built-in RF modem, voice and data communication capabilities; software allows data compression; has 8 A/D inputs
2)	**Interface with DAQ System**	NR
3)	**Major Components**	RF modem, transceiver, antenna
4)	**Transmission Range**	<10 miles
5)	**Transmission Rate**	NR
6)	**Frequency Range**	NR
7)	**Manufacturer/Model**	Dancer Communications, Inc., Messenger 500 Series
8)	**Cost**	NR
9)	**Comments**	Data acquisition plus remote telemetry; wide range of I/O options

Johnson Data Telemetry, Model DL-3412/DL-3422

1)	**Principle of Operation**	RF transmission
2)	**Interface with DAQ System**	NR
3)	**Major Components**	RF modem, transceiver, antenna
4)	**Transmission Range**	10 to 20 miles
5)	**Transmission Rate**	NR
6)	**Frequency Range**	UHF 400 to 500 MHz
7)	**Manufacturer/Model**	Johnson Data Telemetry, Model DL-3412/DL-3422
8)	**Cost**	$645
9)	**Comments**	Meets FCC, European Telecommunications Standards Institute (ETSI), and Canadian regulations; used in SCADA, water management with dataloggers, and other DAQ systems; cost-effective; used in CSI RF telemetry

Adcon Telemetry, Model μT

1) **Principle of Operation**	Analog data input, microcontroller with radio transceiver
2) **Interface with DAQ System**	NR
3) **Major Components**	Radio transceiver, microcontroller interface, antenna, modem
4) **Transmission Range**	1 mile
5) **Transmission Rate**	Up to 19.2 kbaud
6) **Frequency Range**	418 or 433.92 MHz (ETSI) and 902 to 928 MHz (FCC Part 15)
7) **Manufacturer/Model**	Adcon Telemetry, Model μT
8) **Cost**	$1,200
9) **Comments**	Uses an 8-bit microcontroller with 12 kB of one-time programmable applicable storage; designed as a node in a mesh of RF communication devices; four analog input lines

Adcon Telemetry, Model A730MD

1) **Principle of Operation**	Solar-driven module operates over RF in 70-cm telemetry band; can measure analog and digital signals as well as impulses; serial interface offers open connectivity
2) **Interface with DAQ System**	NR
3) **Major Components**	Modem, transceiver, antenna
4) **Transmission Range**	20 km with clear line of sight
5) **Transmission Rate**	1,200 to 19,200 bps
6) **Frequency Range**	420 to 470 MHz radio telemetry band
7) **Manufacturer/Model**	Adcon Telemetry, Model A730MD radio network model
8) **Cost**	NR
9) **Comments**	User programs can be stored in internal EEPROM, enhancing device flexibility; modem meets U.S. and European regulations; data acquisition plus telemetry; optimized for weather monitoring

InterOcean Systems, Inc.

1)	**Principle of Operation**	Spread-spectrum radio modem, cellular, RF, satellite
2)	**Interface with DAQ System**	NR
3)	**Major Components**	System integrator; does not manufacture components
4)	**Transmission Range**	NR
5)	**Transmission Rate**	NR
6)	**Frequency Range**	NR
7)	**Manufacturer/Model**	InterOcean Systems, Inc., Integrated Telemetry
8)	**Cost**	Approximately $15,000 for RF, $25,000 for satellite
9)	**Comments**	Used in oceanography to provide decision-making information for ship captains and operations managers; system integrator; does not manufacture components

duTec, Remote Telemetry Unit

1)	**Principle of Operation**	RF transmitters; spread-spectrum technology
2)	**Interface with DAQ System**	NR
3)	**Major Components**	Depends on telemetry technology
4)	**Transmission Range**	Depends on telemetry technology
5)	**Transmission Rate**	NR
6)	**Frequency Range**	Depends on telemetry technology
7)	**Manufacturer/Model**	duTec, Remote Telemetry Unit
8)	**Cost**	$1,200
9)	**Comments**	The analog modules are the same size as industry-standard digital modules, thus do not require analog-only boards; each module has its own independent, isolated power supply, providing complete electrical isolation; telemetry by spread-spectrum RF, VHF, and phone modems. *An FCC license and a site survey are not required for spread-spectrum RF.*

Campbell Scientific, Inc. (CSI)

1) Principle of Operation	Connects CSI datalogger to host computer via coaxial cable, RF, telephone, cellular phone, or satellite
2) Interface with DAQ System	NR
3) Major Components	Various depending on technology; see text for descriptions
4) Transmission Range	Depends on transmission technology
5) Transmission Rate	NR
6) Frequency Range	Function of transmission technology
7) Manufacturer/Model	Campbell Scientific, Inc.
8) Cost	RF, $2,272; phone, $1,425; cellular, $2,455; satellite, $3,200
9) Comments	Has experience with all forms of telemetry: land-based telephone, cellular, satellite (GOES, Argos, INMARSAT C, Meteosat); telemetry options compatible only with CSI datalogger

7

Vendor List

This appendix provides a listing of some of the vendors found during the literature search described in Appendix A. Products available from these vendors are given as examples of COTS devices/systems in the text and appendix material in this book. This listing is provided solely for informational purposes. Inclusion in the list does not constitute recommendation of the companies or endorsement for use of their products.

Reasonable efforts have been made to ensure that the information provided is up to date. Please note, however, that most toll-free 800 and 888 numbers will function only when dialed within the continental United States.

Acromag, Inc.
30765 South Wixom Road
P.O. Box 437
Wixom, MI 48393-7037
Phone: (248) 624-1541
Fax: (248) 624-9234
E-mail: sales@acromag.com
Website: www.acromag.com

Adcon Telemetry
1001 Yamato Road, Suite 305
Boca Raton, FL 33431
Phone: (561) 989-5309
Fax: (561) 989-5310
E-mail: info@adcon.com
Website: www.adcon.com

Analytical Technology, Inc. (ATI)
680 Hollow Road
Box 879
Oaks, PA 19456
Phone: (610) 917-0991
Fax: (610) 917-0992
Website: www.analyticaltechnology.com

ATI, *see* Analytical Technology, Inc.

Barnant Company
28W092 Commercial Avenue
Barrington, IL 60010-2392
Phone: (800) 637-3739
Fax: (847) 381-7053
Website: www.barnant.com

Campbell Scientific, Inc.
815 West 1800 North
Logan, UT 84321-1784
Phone: (435) 753-2342
Fax: (435) 750-9540
E-mail: info@campbellsci.com
Website: www.campbellsci.com

Capital Controls Co., Inc.
3000 Advance Lane
Colmar, PA 18915
Phone: (215) 997-4000
Fax: (215) 997-4062
E-mail: marketing@capitalcontrols.com
Website: www.capitalcontrols.com

Cole-Parmer Instrument Company
625 East Bunker Court
Vernon Hills, IL 60061
Phone: (847) 549-7600
Fax: (847) 549-7676
E-mail: techinfo@coleparmer.com
Website: www.cole-parmer.com

Component Engineering, Inc.
145 Midgely Drive
Hewlett, NY 11557
Phone: (516) 374-6143
Fax: (516) 374-5327
E-mail: cesco145@ix.netcom.com
Website: www.componentengineering.com

Corning, Inc.
Science Products Division
50 Walnut Street
Corning, NY 14831-0002
Phone: (800) 222-7740
Fax: (607) 974-0345
Website: www.corning.com

CSI, *see* Campbell Scientific, Inc.

Dancer Communications, Inc.
649 South Avenue, Building 6
Secane, PA 19018
Phone: (610) 543-8066
Fax: (610) 543-2151
E-mail: dancer@remotepossibilities.com
Website: www.remotepossibilities.com

Data Electronics U.S.A., Inc.
22961 Triton Way, Suite E
Laguna Hills, CA 92653
Phone: (714) 452-0750
Fax: (714) 452-1170
E-mail: deusa@datataker.com
Website: www.datataker.com

Davis Instruments
4701 Mount Hope Drive
Baltimore, MD 21215
Phone: (800) 368-2516
Fax: (800) 433-9971
E-mail: sales@davisontheweb.com
Website: www.davisontheweb.com

Daytronic Corporation
2211 Arbor Boulevard
Dayton, OH 45439
Phone: (800) 668-4745
Fax: (937) 293-2586
E-mail: sales@daytronic.com
Website: www.daytronic.com

Detora Analytical, Inc.
P.O. Box 2747
Alliance, OH 44601
Phone: (330) 821-6604
Fax: (330) 821-9557
E-mail: DETORAInc@aol.com
Website: www.detora.com

duTec
4801 James McDivitt Road
P.O. Box 964
Jackson, MI 49204
Phone: (517) 750-4700
Fax: (517) 750-4740
E-mail: dutec@aol.com
Website: www.industry.net/dutec

E. Clark & Associates
55 Green Street
Clinton, MA 01510
Phone: (978) 368-3600
Fax: (978) 368-3633
E-mail: sales@eclarkassociates.com
Website: www.eclarkassociates.com

Ellison Sensors, Inc.
1640 Northwest Boca Raton Boulevard
Boca Raton, FL 33432
Phone: (561) 393-0092
Fax: (561) 393-6119
E-mail: bocaesi@cs.com

EMCO
600 Diagonal Highway
Longmont, CO 80501
Phone: (303) 651-0550
Fax: (303) 678-7152
E-mail: marketing@emcoflow.com

Environmental Sensors, Inc.
2759 Pasa Tiempo Glen
Escondido, CA 92025
Phone: (800) 553-3818
Fax: (250) 479-1412
E-mail: esi@esica.com
Website: www.envsens.com

Flowmetrics, Inc.
9201 Independence Avenue
Chatsworth, CA 91311
Phone: (818) 407-3420
Fax: (818) 700-1961
E-mail: information@flowmetrics.com
Website: www.flowmetrics.com

Fluid Metering, Inc.
5 Aerial Way, Ste. 500
Syosset, NY 11791
Phone: (800) 747-5858
(516) 922-6050
Fax: (516) 624-8261
E-mail: pumps@fmipump.com
Website: www.fmipump.com

Fluidyne, Inc.
600 Diagonal Highway
Longmont, CO 80501
Phone: (303) 651-0352
Fax: (303) 678-1754
E-mail: fluidyne@emcoflow.com
Website: www.fluidyne.com

FMI, *see* Fluid Metering, Inc.

GLI International, Inc.
9020 West Dean Road
P.O. Box 23056
Milwaukee, WI 53224
Phone: (800) 454-0263
(414) 355-3601
Fax: (414) 355-8346
E-mail: info@gliint.com
Website: www.gliint.com

Global Water
11257 Coloma Road
Goldriver, CA 95670
Phone: (916) 638-3429
Fax: (916) 638-3270

Great Lakes Instruments, *see* GLI International, Inc.

Hach Company
5600 Lindbergh Drive
Loveland, CO 80538
Phone: (800) 227-4224
(970) 669-3050
Fax: (970) 669-2932
E-mail: csays@hach.com
Website: www.hach.com

Handar, Inc.
1288 Reamwood Avenue
Sunnyvale, CA 94089
Phone: (408) 734-9640
Fax: (408) 734-0655
E-mail: marketing@handar.com
Website: www.handar.com

Hanna Instruments, Inc.
Highland Industrial Park
584 Park East Drive
Woonsocket, RI 02895
Phone: (401) 765-7500
Fax: (401) 765-7575
E-mail: usa@hannainst.com
Website: www.hannainst.com

HI, *see* Hanna Instruments, Inc.

Honeywell, Inc.
MICRO SWITCH Division
Customer Response Center
11 West Spring Street
Freeport, IL 61032
Phone: (800) 537-6945
(815) 235-6847
Fax: (815) 235-6545
E-mail: info@corp.honeywell.com
Website: www.honeywell.com/sensing

Innovative Sensors, Inc.
4745 East Bryson Street
Anaheim, CA 92807
Phone: (714) 779-8781
Fax: (714) 779-9315

INOTEK Technologies, Inc.
11212 Indian Trail
Dallas, TX 75229-3585
Phone: (972) 243-7000
Fax: (972) 243-2924
E-mail: info@inotek.com
Website: www.inotek.com

In-Situ, Inc.
210 South Third Street
P.O. Box I
Laramie, WY 82073-0920
Phone: (307) 742-8213
Fax: (307) 721-7598
Website: www.in-situ.com

InterOcean Systems, Inc.
3540 Aero Court
San Diego, CA 92123-1799
Phone: (619) 565-8400
Fax: (619) 268-9695
E-mail: sales@interoceansystems.com
Website: www.interoceansystems.com

ITT Barton Instrument Systems, LLC
900 South Turnbull Canyon Road
City of Industry, CA 91749
Phone: (626) 961-2547
Fax: (626) 961-4452
E-mail: ittbarton@fluids.ittind.com
Website: www.ittbarton.com

Johnson Data Telemetry Corporation
438 Gateway Boulevard
Burnsville, MN 55337-2559
Phone: (612) 890-8155
Fax: (612) 882-5671

Kari, *see* Scientific Technology, Inc.

KNF Neuberger, Inc.
2 Black Forest Road
Trenton, NJ 08691-1810
Phone: (609) 890-8600
Fax: (609) 890-8323
E-mail: pumps@knf.com
Website: www.knf.com

Kobold Instruments, Inc.
1801 Parkway View Drive
Pittsburgh, PA 15205
Phone: (800) 998-1020
(412) 788-2830
Fax: (412) 788-4890
E-mail: kobold@koboldusa.com
Website: www.koboldusa.com

Logic Beach, Inc.
8363-6F Center Drive
La Mesa, CA 91942
Phone: (619) 698-3300
Fax: (619) 469-8604
E-mail: www-sales@logicbeach.com
Website: www.logicbeach.com

Maihak USA, Inc.
475 South Main Street
Jasper, GA 30143
Phone: (706) 692-6949
Fax: (706) 692-6131
E-mail: maihakusa@aol.com
Website: www.maihak.de

Manostat, *see* VWR

MC Meters and Controls Company
9244 Compton Square Drive
Cincinnati, OH 45231
Phone: (513) 931-5555
Fax: (513) 931-5558
E-mail: meters5@aol.com

Met One, Inc.
481 California Avenue
Grants Pass, OR 97526
Phone: (800) 866-7889
(541) 479-1248
Fax: (541) 479-3057
E-mail: metone@cdsnet.net
Website: www.metoneinc.com

Modus Instruments, Inc.
55 Green Street
Clinton, MA 01510
Phone: (978) 365-9300
Fax: (978) 368-3633
E-mail: sales@eclarkassociates.com

MONITEK Technologies, Inc.
1495 Zephyr Avenue
Hayward, CA 94544
Phone: (800) 458-4454
(510) 471-8300
Fax: (510) 471-8647
E-mail: sales@monitek.com
Website: www.monitek.com

Noshok, Inc.
17820 Englewood Drive
Middleburg Heights, OH 44130-3445
Phone: (440) 243-0888
Fax: (440) 243-3472
E-mail: noshok@noshok.com
Website: www.noshok.com

OMEGA Engineering, Inc.
1 Omega Drive
P.O. Box 4047
Stamford, CT 06907-0047
Phone: (800) 826-6342
(203) 359-1660
Fax: (203) 359-7700
E-mail: info@omega.com
Website: www.omega.com

Onset Computer Corporation
P.O. Box 3450
Pocasset, MA 02559-3450
Phone: (508) 759-9500
Fax: (508) 759-9100
E-mail: sales@onsetcomp.com
Website: www.onsetcomp.com

Opto 22
43044 Business Park Drive
Temecula, CA 92590-3614
Phone: (800) 321-6786
(909) 695-3000
Fax: (800) 832-6786
(909) 695-3095
E-mail: sales@opto22.com
support@opto22.com
Website: www.opto22.com

Orion Research, Inc.
500 Cummings Center
Beverly, MA 01915-6142
Phone: (800) 2251480
(978) 232-6000
Fax: (978) 232-6015
E-mail: domcs1@orionres.com
intcs1@orionres.com
Website: www.orionres.com

Rosemount Analytical, Inc.
Uniloc Division
2400 Barranca Parkway
Irvine, CA 92606
Phone: (714) 863-1181
Fax: (714) 863-9159
Website: www.rauniloc.com

Scientific Technology, Inc.
6550 Dumbarton Circle
Freemont, CA 94555
Phone: (888) 510-4357
(510) 608-3400
Fax: (510) 744-1442
E-mail: sales@sti.com
Website: www.sti.com

Setra Systems, Inc.
159 Swanson Road
Boxborough, MA 01719-1304
Phone: (800) 257-3872
(978) 263-1400
Fax: (978) 264-0292
E-mail: sales@setra.com
Website: www.setra.com

Sievers Instruments, Inc.
6060 Spine Road
Boulder, CO 80301
Phone: (303) 444-2009
Fax: (303) 444-9543
E-mail: sievers@sieversinst.com
Website: www.sieversinst.com

Solomat
The Waterside Building
26 Pearl Street
Norwalk, CT 06850
Phone: (800) 765-6628
(203) 849-3111
Fax: (203) 847-9320
E-mail: iaq@solomat.com
Website: www.solomat.com

Synetcom Digital, Inc.
1426 Aviation Boulevard, Suite 203
Redondo Beach, CA 90278
Phone: (310) 379-2000
Fax: (310) 372-2331
E-mail: info@synetcom.com
Website: www.synetcom.com

Tytronics, Inc.
25 Wiggins Avenue
Bedford, MA 01730-2323
Phone: (781) 275-9660
Fax: (781) 275-9665
E-mail: info@tytronics.com
Website: www.tytronics.com

VWR
1310 Goshen Parkway
West Chester, PA 19380
Phone: (800) 932-5000
(610) 431-1700
Fax: (610) 429-9340
E-mail: solutions@vwr.com
Website: www.vwrsp.com

Wedgewood Technology, Inc.
300 Industrial Road
San Carlos, CA 94070
Phone: (800) 241-8404
(650) 593-1598
Fax: (650) 593-0235
E-mail: wedgewood@wedgewoodtech.com
Website: www.wedgewoodtech.com

Index

D

E

R

S

T

U

V

W

X